ENGINEERING GRAPHICS

A Problem-Solving Approach

D. McAdam
R. Winn

Addison-Wesley

An imprint of Addison Wesley Longman Ltd.

Don Mills, Ontario • Reading, Massachusetts • Harlow, England
Melbourne, Australia • Amsterdam, The Netherlands • Bonn, Germany

Publisher: Ron Doleman
Managing Editor: Linda Scott
Editor: Tusker International Consulting Ltd.
Designer, Desktopping, and Cover Design: Anthony Leung
Production and Manufacturing Coordinator: Alexandra Odulak

Canadian Cataloguing in Publication Data

McAdam, D. (Donald), 1940–
    Engineering graphics: a problem-solving approach

Includes index.
ISBN 0-201-89200-6

1. Engineering graphics. 2. Engineering graphics—Problems, exercises, etc.
I. Winn, R. (Roger).   II. Title.

T353.M32 1999   604.2   C99-931320-7

ISBN 0-201-89200-6

Printed and bound in Canada.

A B C D E -WC- 03 02 01 00 99

Contents

PREFACE

Engineering Graphics: A Problem-Solving Approach is an innovative text that provides a fresh perspective to engineering graphics. It is designed for engineering students taking a one-semester engineering graphics course.

A first-year engineering program is often full of mathematics, chemistry, and physics courses. An engineering graphics course may be the only engineering course in that year. In many instances, students have little knowledge of materials, manufacturing, and mechanics. Often they have no background in engineering and they are not equipped to do any engineering calculations (although calculations make up only a very small part of engineering work). This is where *Engineering Graphics: A Problem-Solving Approach* fills an important gap.

The goal of *Engineering Graphics: A Problem-Solving Approach* is to engage students in critical thinking about graphics problems. It is also a quick source of information that many students use to succeed in their engineering courses and careers. It is *not* intended to train drafters in all aspects of engineering drawing.

Engineers must know how to read engineering drawings and understand the methods and conventions used in creating them. They must be able to visualize three-dimensional objects from two-dimensional drawings; explain their ideas and plan their work with sketches; apply what they know to new problems; approach projects in an organized, systematic way; and communicate their findings in an effective manner. *Engineering Graphics: A Problem-Solving Approach* is intended to develop all of these skills.

Engineering Graphics: A Problem-Solving Approach is divided into eight chapters that cover the principles of projection, sketching, sectioning, dimensioning, engineering drawings, visualization, intersection and development, and presenting technical information. It also has a series of appendices that provide a wealth of useful supplementary information such as vector analysis, standards, symbols, scales, and notations.

Engineering Graphics: A Problem-Solving Approach contains a brief introduction to tolerances but does not cover geometric tolerances. Some students will require this in their later studies, especially in manufacturing courses; however, many students will never need more than a basic understanding of tolerances.

Each chapter of *Engineering Graphics: A Problem-Solving Approach* presents information in an easy-to-read format with plenty of illustrations. To assist students who may know little about materials, manufacturing, and machining processes, *Engineering Graphics: A Problem-Solving Approach* uses practical examples and a step-by-step approach to explain a variety of engineering drawings and calculations.

In addition, *Engineering Graphics: A Problem-Solving Approach* has a large number of graphics problems in each chapter that challenge students to apply their knowledge. The problems range from basic ones that establish familiarity with the chapter content to challenging, open-ended ones.

Open-ended problems, with several correct solutions, are the norm in engineering, but traditionally students do not encounter them until their later years of study. Open-ended problems are frustrating to many students who are used to the "one correct answer problem." These problems are included in *Engineering Graphics: A Problem-Solving Approach* to encourage students to think on their own and to build confidence in their own abilities without an answer to check. In addition, these problems help students to develop a systematic approach to problem solving that will be invaluable to them in their future engineering studies and careers.

Chapter 1 of *Engineering Graphics: A Problem-Solving Approach* introduces students to the principles of projection. This information forms the foundation for the following chapter that introduces students to sketching.

Chapters 3 and 4 of *Engineering Graphics: A Problem-Solving Approach* on sectioning and dimensioning introduce students to basic techniques, standard conventions, and guidelines. Because sectioning and dimensioning are not random processes, the techniques, conventions, and guidelines presented in the chapters help students to detail simple parts correctly.

Since first-year engineering courses include students who will go into different branches of engineering, and because there is so much interaction between branches of engineering, Chapter 5 introduces students to drawings from mechanical, civil, chemical, electrical, and electronic engineering.

It is not possible to cover all aspects of engineering drawings in a one-semester course; however, some of the most common ones have been included in Chapter 5. Students who need to learn more about some topic in later years will have a sufficiently good basic background to do so after completing Chapter 5.

In Chapter 6 of *Engineering Graphics: A Problem-Solving Approach*, students learn about visualization, with particular emphasis on points, lines, and planes. This chapter also includes information on contours and 3-D representation.

A computer is merely a tool to help users be more productive. While many engineering drawings are created with computer programs, such programs do not think for the user. People who do not know anything about engineering drawings, rules, and conventions, find a computer program of little use to them. Just because a person draws something using a computer does not make it correct. The computer adage: "Garbage in, garbage out" always applies. Users need to know what they want to do before they can use their computer effectively. To underscore this point, the last section of Chapter 6 shows students how to solve various problems using a computer.

Chapter 7 of *Engineering Graphics: A Problem-Solving Approach* introduces students to intersection and development. This chapter includes a large number of practical examples to help students thoroughly understand the concepts presented in the chapter.

Drawings are one form of engineering communication but technical writing is even more important. Engineers spend 60 - 70 percent of their time communicating in one form or another. They need to be able to answer questions and convey their ideas in an understandable manner. Few people will advance very far in engineering if they cannot express their ideas in a clear, concise, concrete, correct, and courteous manner (the five Cs of communication).

Many employers complain that engineering graduates cannot communicate, so Chapter 8 of *Engineering Graphics: A Problem-Solving Approach* introduces students to the basics of technical writing and provides hints on how to prepare effective communications such as business letters and engineering reports.

This book does not include graphical calculation techniques because these are available in spreadsheet and mathematical analysis programs.

Acknowledgments

The publishers and authors would like to thank the following people who reviewed the manuscript at some stage in its development: David Bonham, Richard Burton, Bruce Dunwoody, G. Glinka, Mahmoud Haddara, Kefu Liu, Robert Marshall, Fred Spinney, Roger Toogood, Allan Torvi, and Paul Zsombor-Murray. A number of people contributed to this book, most notably Aaron Bohnen, who drew many of the drawings for problems and figures, and Parsa Pirseydi, who was responsible from some of the drawings for problems.

Thanks to Ron Doleman, our publisher at Addison Wesley Longman, and to Linda Scott, the managing editor, for their encouragement and suggestions. Our editor, Jennie Bedford, was very helpful, and her editorial skills and suggestions were appreciated.

I would especially like to thank my wife, Karen, for her patience, understanding, and support throughout the creation of this book; I dedicate this book to her. (D.M.)

Many thanks to my family for their support while I was working on this book. (R.W.)

Chapter 1

Principles of Projection

When I was a kid growing up in Far Rockaway, I had a friend named Bernie Walker. We both had "labs" at home, and we would do various "experiments." One time we were discussing something—we must have been eleven or twelve at the time—and I said, "But thinking is nothing but talking to yourself inside."

"Oh yeah?" Bernie said. "Do you know the crazy shape of a crankshaft in a car?"

"Yeah, what of it?"

"Good. Now, tell me: how did you describe it when you were talking to yourself?"

So I learned from Bernie that thoughts can be visual as well as verbal.

Richard P. Feynman, Nobel Laureate, Physics, "What Do You Care What Other People Think?" New York: W.W. Norton and Company Inc., 1988.

The world we live in is three dimensional. We are surrounded by three-dimensional objects. We are also surrounded by two-dimensional representations of these objects in pictures on a television screen or the pages of a book. In our minds, we translate these pictures into three-dimensional objects. With familiar, everyday objects, this happens automatically. But what if the picture is of something you have not seen before? What if you have a picture in your "mind's eye" that you want to share with someone else? Imagine trying to explain how to make a printed circuit board without drawing the circuit, or how to construct a building, or an automobile, without drawing a picture. Could you describe a safety pin?

As personal computers become faster and more powerful and graphics capability improves, more and more information can be presented in graphical form. The ability to create two-dimensional images representing three-dimensional space is becoming more important. As an engineer, you will need to show your ideas to other members of the design and development team. Your ideas are of little value if you cannot communicate them.

Engineers spend 60-70 percent of their time communicating with drawings, pictures, and written and verbal presentations. Like all professions, engineering has its own language and conventions. If you are going to communicate your ideas to the engineering world, you must learn to speak the language.

The universal language of engineering is drawing. How else could the information required to build an airplane or construct an office building be conveyed to the people who build them? You may not be able to speak the language of the country in which you are doing an engineering project, but you can understand the drawings. Conventions and special methods of presentation are used so that engineering drawings are easy to create and understand. A particular feature may not be drawn as it would actually appear to the eye, but in a simpler way which is easier and faster to draw. These conventional methods will be discussed in later chapters. You will eventually become so familiar with them that you will be able to understand them immediately.

Let's start by looking at how objects are formed.

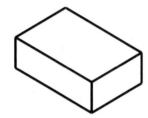

Figure 1.1
Rectangular prism

How Objects Are Formed

Objects are formed by combining simple, basic shapes (called **primitives**). A rectangular prism, shown in Figure 1.1, is an example of a basic shape.

These basic shapes can be enlarged, shrunk, combined, or subtracted from other basic shapes to create new, more complicated objects. For example, if a small rectangular prism is removed from a larger rectangular prism, a new object is created (see Figure 1.2).

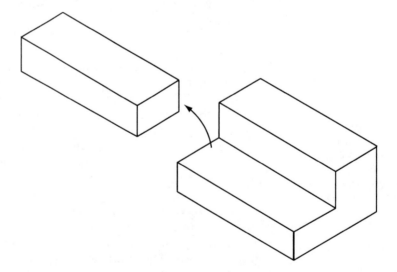

Figure 1.2   Removing one basic shape from another to create a new object

The new object, shown in Figure 1.3, is more complex because it has more surfaces, but it is still made up of simple shapes. The same shape could be created by combining three rectangular prisms.

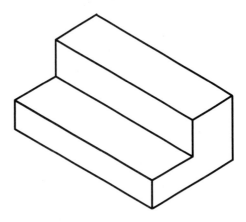

Figure 1.3   New object resulting from removing a rectangular prism

Obviously, a more complex object can be created by combining as many basic shapes as required. A cylinder, triangular prism, and rectangular prism, shown in Figure 1.4, can be combined to form a simple building with a chimney, as shown in Figure 1.5.

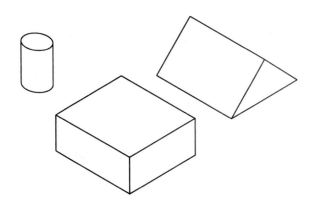

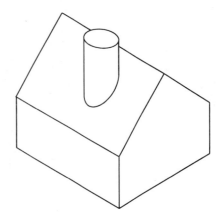

Figure 1.4
Three basic shapes: cylinder, triangular and rectangular prisms

Figure 1.5
Object created from cylinder, triangular and rectangular prisms

You must exercise judgment when combining basic shapes, otherwise you might draw objects that cannot exist. For example, removing two small rectangular prisms from opposite corners of a larger prism can result in an object that cannot exist, as illustrated in Figure 1.6. The connection between the two parts is a line that has only one dimension. It has no strength and the object does not stay together.

CONNECTION HAS NO THICKNESS

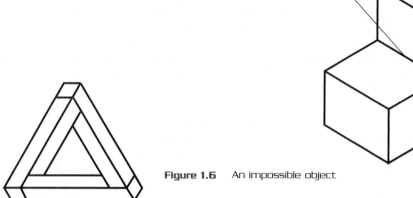

Figure 1.6   An impossible object

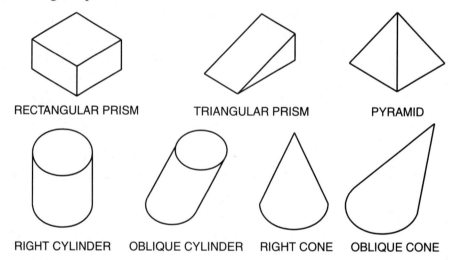

Figure 1.7
An impossible object constructed of three square bars

Similarly, it is possible to combine three square prisms to form an impossible object, which can be drawn in two dimensions, but which cannot be constructed in three dimensions. Figure 1.7 is an example of an impossible object constructed from three square bars. This "object" can be drawn but cannot be made.

More basic shapes are shown in Figure 1.8. A sphere and cube are not shown. A sphere is simply a circle and a cube is simply a special case of a rectangular prism.

RECTANGULAR PRISM   TRIANGULAR PRISM   PYRAMID

RIGHT CYLINDER   OBLIQUE CYLINDER   RIGHT CONE   OBLIQUE CONE

Figure 1.8   Some basic shapes

Complex shapes, that require the combination of many basic shapes, can be quickly created using a computer. The process of forming a solid by combining basic shapes is called **solid modeling**. A solid model is a pictorial representation of an object and is the easiest way to visualize it. Pictorial representations give the illusion of three dimensions, but they are not used for fabrication. Pictorial drawings are covered later in this chapter. Engineering drawings use another method of showing three-dimensional objects.

Before drawing objects, you need to understand the concept of orthographic projection.

Orthographic Projection

Three things must be defined before an object can be drawn: the location of the eye, or viewpoint; the object; and the picture plane, on which the object is drawn. The arrangement of these three elements is called **orthographic projection**.

If you were to put the object to be drawn behind a piece of glass and look at it through the glass, a line drawn from your eye to a point on the object would pass through the glass at some point. This point represents a point on the object. The glass represents a picture plane on which the object could be drawn. If the viewer were located an infinite distance from the object, the line from the eye would pass through the glass picture plane at 90° and all lines from the eye to points on the object would be parallel. This arrangement is shown in Figure 1.9.

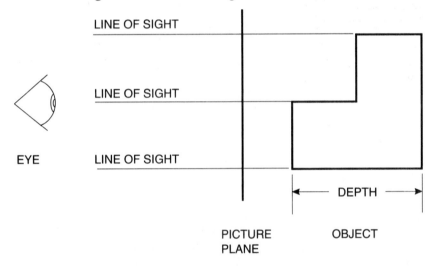

Figure 1.9   Placement of object, eye, and picture plane

A picture plane is located between the eye and the object, and lines of sight are drawn from the object to the eye (at infinity), as illustrated in Figure 1.10.

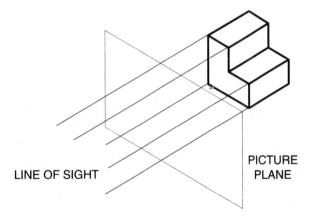

Figure 1.10   Lines of sight from the eye to points on the object

If the intersections of lines of sight with the picture plane are connected, as in Figure 1.11, the outline of the object and features that can be seen from the chosen viewpoint are seen on the picture plane.

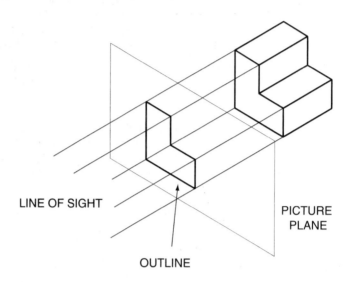

LINE OF SIGHT

PICTURE PLANE

OUTLINE

Figure 1.11 Joining points where lines of sight intersect picture plane

Figure 1.12 shows a view of the object as it appears on the picture plane, which could be a computer screen or a piece of paper, and the object is behind it (inside the computer).

This arrangement of the eye, object, and picture plane is called *orthographic projection* because the lines of sight intersect the picture plane at right angles. *Orthogonal* means "at right angles to something else." This type of orthographic projection is used in North America. A different type of orthographic projection, one used in Europe and other parts of the world, is described later.

Obviously, Figure 1.12 does not define the object entirely, so more information is required to show what it looks like. For example, there is no way to see the depth in Figure 1.12. We must view the object from other viewpoints before we can determine the shape. This requires more picture planes and different viewpoints.

Viewpoints and picture planes can be located anywhere, but there are six principal picture planes that are always in the same relative position. The six principal planes can be thought of as the six sides of a cube, each side representing a picture plane.

Figure 1.13 shows a cube with three sides labeled: front, top, and right side. The other three sides (left side, back, and bottom) cannot be seen. The front, top, and right side correspond to the side of the object that is seen on that face of the cube. These three sides are called **principal planes**.

Usually only three planes are required to define an object.

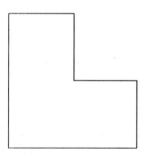

Figure 1.12
Projection on the picture plane

Figure 1.13
Three principal planes

Imagine that the object shown in Figure 1.11 is put in the center of a glass cube. Each side of this cube represents a plane on which the object will be projected.

The object is projected onto these planes by joining the points where the lines of sight intersect the plane. The result is shown in Figure 1.14, where the front, top, and right-side views are seen on the front, top, and right-side planes. Not all lines of sight are shown in this figure.

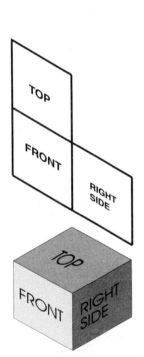

Figure 1.15
Orientation of views

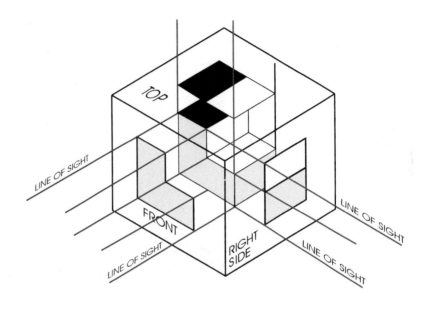

Figure 1.14   Forming orthographic views

This shows how orthographic views are formed, but we need to represent an object on a two-dimensional surface, not on the sides of a glass box. To do this, the top and the right-side views are folded so that they are in the frontal plane, that is, the same plane as the front view. The top is folded up and the right side is folded out so that both are in the frontal plane. This unfolding of the top and right side is shown in Figure 1.15.

The three views of the object are shown in Figure 1.16.

The views are named here, but they need not be. Because these views are always in the same positions relative to each other, we know which views they are by their position. The length can now be seen in either the top view or the right-side view and, of course, it must be the same in each view. Note that there is correspondence between the views. The object does not move or change in size; it is simply seen from a different viewpoint.

While there are six sides to a cube, only three have been shown. The three sides that cannot be seen in Figure 1.15 are unfolded into the frontal plane. Figure 1.17 shows the cube with the back, left-side, and bottom faces partially unfolded so they are visible. The top and right sides have not been unfolded in Figure 1.17.

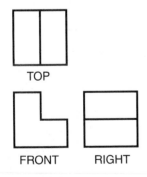

Figure 1.16
Three orthographic views

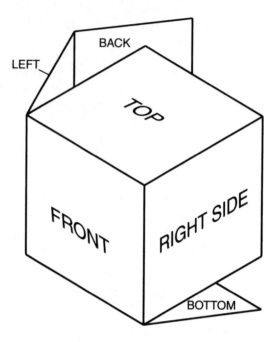

Figure 1.17   Back, left-side, and bottom planes partially unfolded

The result of unfolding all six sides of the cube is shown in Figure 1.18. Six views are available, and all sides of the object could be shown if required.

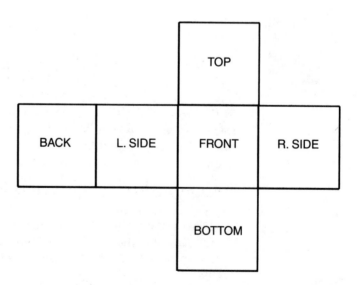

Figure 1.18   Position of six sides of a cube when folded into the frontal plane

All six views are shown in Figure 1.19, although it would be very unusual to require all six views to define an object.

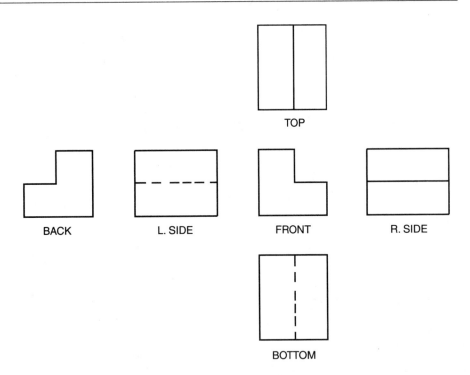

Figure 1.19   Six orthographic views

In this case, the left-side view shows very little. The dashed line indicates that the line is hidden; however, it can be seen in the right-side view. The bottom view gives the same information as the left-side view, but this is seen better in the top view. The back view is a mirror image of the front view. This object can be completely described with three views—front, top, and right side—and there is no need to draw the other three views.

The three most commonly used views are the front (there is always a front view), top (also called a **plan view**), and the right-side view. A left-side view would be used instead of a right-side view if a feature could only be seen in the left-side view. Some objects may require both right- and left-side views. Back or bottom views are seldom required.

Views other than the six principal ones can be used if required. These other views, called **auxiliary views**, are dealt with in greater detail later.

Now let's look at how to choose front and side views.

Choosing Front and Side Views

Before drawing anything, you must decide how it should be oriented so that maximum information is given in the front view. The front view should show the shape of the object and there should be a minimum of hidden lines. You must also think about the top and side views when deciding on the front view, since these are projected from the front view. The object should appear in the front view as it would normally be used. Sometimes this is not known, but usually it is obvious. It would not make much sense to show an automobile upside down with the tires in the air, since automobiles are not normally used in this position.

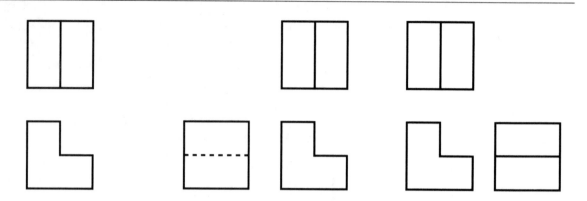

Figure 1.20
Two orthographic views: top and front

Figure 1.21
Orthographic views showing the left-side, top, and front views

Figure 1.22
Orthographic views showing the top, front, and right-side views

The front view of the object shown in Figure 1.20 is chosen to show the characteristic shape (an "ell" shape). The corresponding top view shows the length. When two views are given in this orientation, the lower view is the front view. Since views always have the same locations relative to each other, there is no need to name them, unless there is a possibility of misunderstanding. If this is the case, identify at least one view.

This simple object can be defined with two views, but, to avoid misunderstanding, it is usual to draw three views. Either a left-side view or a right-side view will be added. As only one will be used, be sure to select the view that best shows details of the object. The shape of the object must be considered to determine which is the best choice in any particular case. Figure 1.21 shows how the three views would appear if a left-side view were chosen.

The middle edge of the "ell" shape is not seen in the left-side view since it is on the back-side and is hidden. This is represented by the dashed line across the middle of the shape. This dashed line represents the edge of the "ell." Because it is behind another surface, it is called a **hidden line**, so instead of choosing the left-side view, it is better to choose the right-side view. The object then appears as shown in Figure 1.22.

The middle surface appears as a solid line in the right-side view, so the right-side view is a better choice than the left-side view because it gives a better picture of the object. This is a very simple object and so the choice is easy to make. Views selected to represent a more complex object would be selected using the same criteria.

The side view chosen depends on the object. If details are on the left-side after the best front view is chosen, the left side would be drawn. Right-side views are more common; however, there is no rule that this must be the case. Sometimes it is necessary to draw both right- and left-side views to fully define an object. If it does not matter whether a left-side view or a right-side view would define the object, a right-side view is usually used.

Another way of positioning the "ell"-shaped object is shown in Figure 1.23. The front view shows the shape; however, the corresponding top view shows one edge as a hidden line.

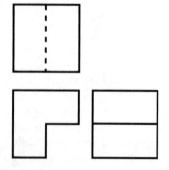

Figure 1.23
Orthographic views showing a poor choice of front view

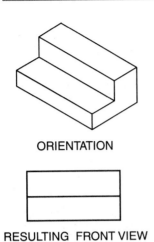

ORIENTATION

RESULTING FRONT VIEW

Figure 1.24
Object orientation and corresponding front view

The front view satisfies the criterion of showing the shape, but the top view is not as illustrative as it could be. This can easily be improved by positioning the object as shown in Figure 1.22. Another possible orientation is shown in Figure 1.24.

This orientation also results in a poor front view that shows nothing of the "ell" shape. There are several ways this front view could be interpreted. The front view is the most important view, but it is only one of several views and must be considered as part of a group that defines the object in the best possible way.

Now let's look at another method of orthographic projection.

Another Method of Orthographic Projection

A different arrangement of object, picture plane, and eye is used in Europe. The picture plane is positioned to the right of the object so the order becomes eye, object, and picture plane. This arrangement, called **first angle projection**, is illustrated in Figure 1.25. The same front view appears on the picture plane with each method of projection.

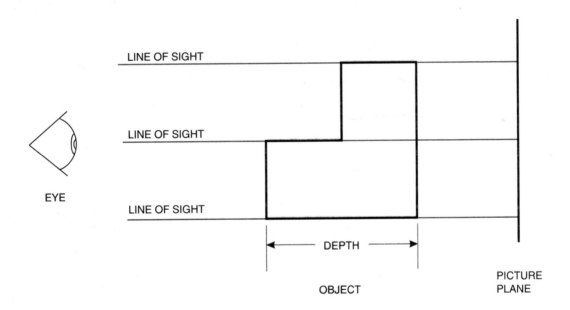

Figure 1.25   First angle projection

The representation on the picture plane is the same for each method of projection. A truncated cone is used as a symbol to indicate which method of projection is used (Figure 1.26). Each projection method illustrated in Figure 1.26 shows the same information, but the relative positions of the object and picture plane are different. The appropriate symbol is usually located in the lower right corner, in, or near, the title block to show which method of projection is used.

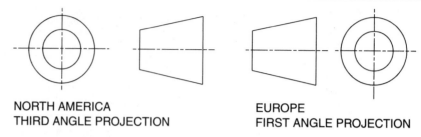

NORTH AMERICA
THIRD ANGLE PROJECTION

EUROPE
FIRST ANGLE PROJECTION

Figure 1.26   Symbol used to show which method is used

The method of projection used in North America is called **third angle projection** and in Europe, **first angle projection.**

Now let's move on to look at how to project between orthographic views.

Projection Between Orthographic Views

We have seen how orthographic views can be created from a solid by projecting from points on the solid to projection planes, but we must also know how to project between different orthographic views. Given two views of any point, the same point in any other orthographic view can be found. The corners of one face of the solid shown in Figure 1.27 have been identified *abcdef* in the side view. The same points are identified in the top view. Point *a* is above *f* and *d* is above *e*. The front view face *abcdef* can be drawn by projecting these points from the side and top views. Point *a*, in the front view, must be at the intersection of the projection line from the side view and the top view. Point *b* must be at the point where projections from the side view and top view intersect.

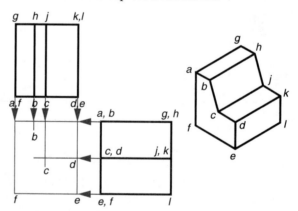

Figure 1.27   Locating points in a third view

The complete face can be drawn by projecting all points into the front view. The points at the ends of the lines making up the surface must be joined in consecutive order. They cannot be connected in the order *abdcef*, for example. There are no lines *bd* or *ce*. The technique of labeling points is useful when determining what a more complex object looks like.

If you are wondering why the label "*i*" has not been used, it is to prevent confusion with "1" on drawings.

Now let's look at how curved surfaces are shown.

Curved Surfaces

A flat, or plane, surface can be defined by straight lines around the perimeter. A curved surface can also be defined in this way but there may be no indication that the surface is curved. To illustrate this, look at Figure 1.28. Figure 1.28 shows a rectangle. The thin, broken line through the middle of the rectangle is a centerline and is not part of the object. This is a view of a cylinder, but there is little to indicate that the surface is curved. The centerline is the only clue. From this example you will see how important it is to use more than one view to define an object with curved surfaces.

Two views of the cylinder are shown in Figure 1.29. The circular shape is seen in the side view.

One thing common to almost all curved surfaces is a centerline. The symbol for a centerline is shown in Figure 1.30.

Figure 1.28
One view of a cylinder

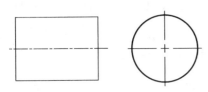

Figure 1.29
Two views of a cylinder

Figure 1.30   Centerline symbol

A centerline symbol is a sequence of long and short lines. If a centerline is used to indicate the center of a cylinder, for example, the center point is indicated by the intersection of the short dashes as shown in Figure 1.29. A centerline is used to indicate

a) the center of a circle,
b) the center of a circular arc,
c) an axis of symmetry.

These uses are shown in Figure 1.31. A centerline extends 2-3 mm beyond the item to which it refers and does not extend between views (Figure 1.29). A curved surface is usually confirmed by giving two views of the object.

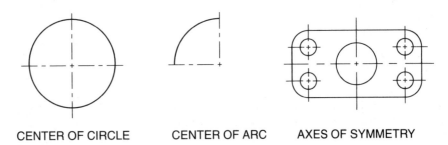

CENTER OF CIRCLE   CENTER OF ARC   AXES OF SYMMETRY

Figure 1.31   Uses of a centerline

Now that you have a basic understanding of orthographic projection, let's move on to look at pictorial representation.

Pictorial Representation

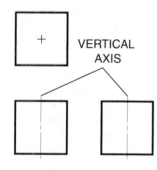

Figure 1.32
Three views of a cube

A pictorial drawing has the illusion of three dimensions, shows more information, and is often easier to visualize, particularly by persons with no engineering background. Pictorial drawings are used as sketches to explore ideas, aid explanation, and present information. A pictorial is often used to show the relationship between different parts of an assembly and how it is put together. Pictorial drawings are not used for manufacture or construction.

Isometrics

There are several types of pictorial drawing; however, the most commonly used in engineering is the **isometric**, which is explained here. Other types have advantages in specific applications and are discussed later. The pictorial drawings used in this book are isometrics. A cube, shown in Figure 1.32, will be used to illustrate how an isometric is created.

Figure 1.32 shows three orthographic views—top, front, and right side—of a cube. Each face appears as a square and only one side is seen in each view. The reason for drawing a pictorial is to show more sides of the object in one view.

We have seen that different views result from different viewpoints; however, similar results can be obtained by repositioning the cube. First, the cube is rotated 45° about a vertical axis through the center of the cube. Two sides, instead of one, can be seen in the front view (Figure 1.33); however, these sides appear shorter than they actually are. All three views of the cube after rotation are shown in Figure 1.33.

If the cube is now tipped forward about the corner "X," the top surface can be seen in the front view. Figure 1.34 shows the front view and corresponding side and top views after tipping. Two sides and the top can now

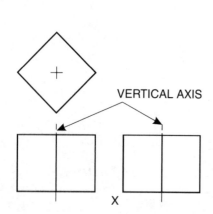

Figure 1.33
Views after rotation about vertical axis

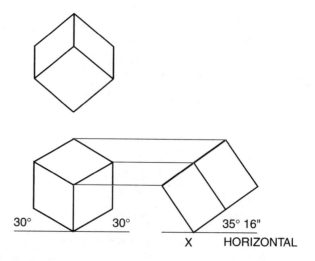

Figure 1.34
Three orthographic views after tipping about "X"

be seen in one view, the front view, and there is no need to draw the side or top views. They are shown here to illustrate how an isometric *projection* is created. It is simply an orthographic drawing with the object oriented in a particular position.

The angle through which the cube is tipped is chosen so that all sides are equal length and make an angle of 30° to the horizontal. *Isometric* comes from a Greek word meaning "equality of measure." The sides are shortened to 82 percent of actual length, but this shortening is neglected in making an isometric *drawing* and all sides are drawn actual size. This results in a small distortion which is noticeable only in very large objects.

A set of isometric axes is shown in Figure 1.35. Computer graphics programs often have the facility to create a grid of 30° lines on the screen as an aid to creating an isometric. Special paper with a 30° grid is available for sketching.

A cube drawn as an isometric appears as shown in Figure 1.36. All sides are equal and are drawn full size. Any line parallel to an isometric axis can be measured as being true length. A line which is not parallel to an axis is not true length, as illustrated by the diagonals on the vertical sides.

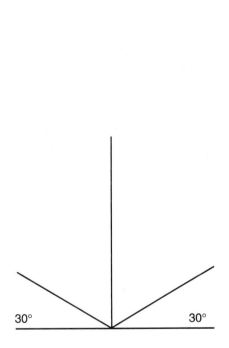

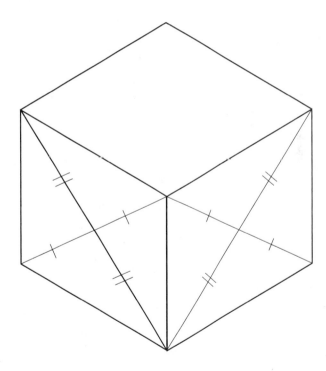

Figure 1.35   Isometric axes

Figure 1.36   Isometric drawing of a cube

All diagonals on a cube are equal length but it is obvious that they are not equal on the drawing.

Now let's explore some other pictorials.

Other Pictorials

An isometric drawing is created by first placing an object in a specific position and drawing a front view as if it were full size. The position is chosen to give equal lengths on each of the axes. If different angles are used in positioning the object, a different pictorial is obtained. The object can be oriented to give the same scale on either none, two, or three axes. These different representations are called **isometric**, **dimetric**, and **trimetric**. Axes for the three types are shown in Figure 1.37. The general term for this type of drawing is an **axonometric**.

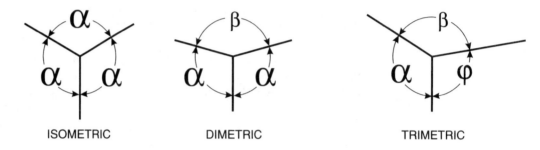

ISOMETRIC   DIMETRIC   TRIMETRIC

Figure 1.37   Comparison of isometric, dimetric, and trimetric axes

Angles between axes on an isometric are the same, so the scale is the same on each axis. A dimetric drawing has equal scales on two axes and a trimetric has a different scale on each axis. These different scales make dimetric and trimetric more difficult and time-consuming to draw, therefore isometric is the most commonly used. Isometric drawings are used for pictorials in this book.

Some computer programs will generate an isometric from given orthographic drawings and some will allow an object to be rotated and turned so it can be viewed from any angle.

Sometimes isometric pictorials are not suitable so another type of pictorial, oblique projection, is used.

Oblique Projection

There are times when an object has features that may be difficult, or time-consuming, to represent with an isometric pictorial. In such instances, a different type of pictorial may be quicker and easier to draw. For example, a circular feature appears elliptical in an isometric and an ellipse takes longer to draw than a circle. While computer programs will draw ellipses, it is generally easier and faster to position and draw a circle. It is in circumstances such as this that a different type of projection, called oblique projection, can be used to advantage.

An orthographic projection results if lines of sight from the eye to the object pass through the picture plane at 90°. If lines of sight pass through the picture plane at an angle other than 90°, the result is called an **oblique projection**. Figure 1.38 shows a comparison of the arrangement for orthographic and oblique projections. Lines of sight are parallel in both types of projection.

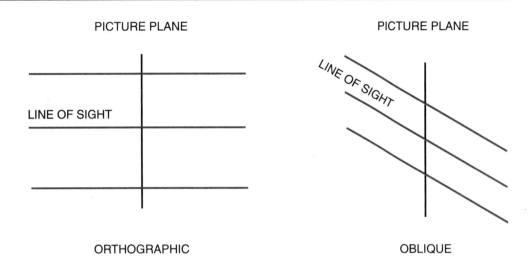

Figure 1.38   Comparison of orthographic and oblique projection

The front view is always drawn full size in an oblique drawing. Features in the front view are seen in their true shape and circular features are easier to draw. The disadvantage is that the side in the direction of the receding axis can appear distorted.

Although lines of sight can theoretically pass through the picture plane at any angle, only a few angles are used in practice. The most commonly used angles are 30°, 45°, and 60°.

Two types of oblique projection, called **cavalier projection** and **cabinet projection**, are in common use. These are similar, but distortion on the receding axis is reduced with cabinet projection. Figure 1.39 compares cavalier and cabinet drawings.

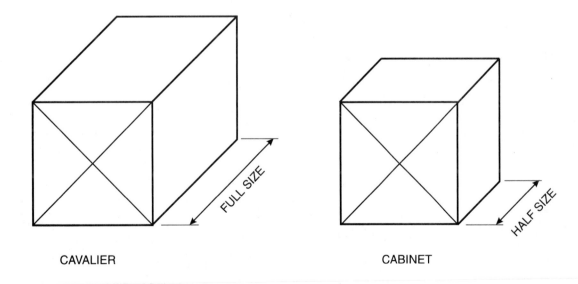

Figure 1.39   Comparison of cavalier and cabinet drawings

As you can see, the front view is full size in both types, but the scale on the receding axis is different. A cavalier drawing is drawn full size on all axes. If the object is long in the direction of the receding axis, there is distortion and the object does "not look right." A cabinet drawing is drawn half size on the receding axis to compensate for this. The difference can be seen even in these simple drawings. Distortion is reduced by changing the scale on the receding axis. Although this process takes more effort, it is not difficult. Lines in the direction of the receding axis can be drawn full size and easily scaled down with a computer drawing program.

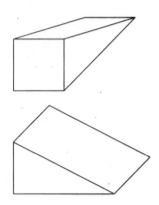

Figure 1.40 Comparison of two front views with an oblique projection

Oblique drawings are easier and faster in some applications. Since the front view is drawn full size, it is easier to draw than if it is scaled up or down, and this can be used to advantage. The features of the object must be shown in the most descriptive way and the drawing must be easy to do. Curved features are less work to draw full size and the front view can be chosen with this in mind, but there is another consideration in choosing the front view. (There usually are other considerations in engineering decisions.) If the object is longer in one direction than another, the longest side should be shown as the front view to minimize distortion. Figure 1.40 shows a comparison of two possible front views with an oblique projection.

This results in two possible front views, depending on the object, so you need to decide which one to use. You must bear in mind the objectives of the pictorial. Shape and details of the object must be shown clearly with a minimum of effort. In this case, the front view would show the curves since it is easier to draw these in full size. The distortion on the receding axis would be secondary in this case.

Figure 1.41 shows a cavalier and cabinet drawing of an object which is represented well with an oblique drawing. The front view shows the true shape. Distortion on the receding axis is more obvious with the cavalier drawing, so the cabinet drawing is the best pictorial to use for this object.

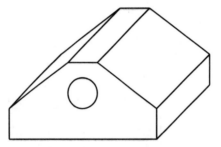
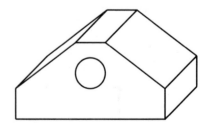

Figure 1.41   Cavalier and cabinet drawings of the same object

The angle used in Figure 1.41 is 45°; however, other angles could be chosen to show the object in the most descriptive way. The receding scale can also be chosen to minimize distortion and show the best representation. It is easily rescaled using the "scale" commands on your computer program.

Yet another type of pictorial is the perspective drawing. Let's look at this now.

Perspective Drawings

A **perspective drawing** represents what the eye sees. You are all familiar with the view looking along railway tracks: The tracks seem to come together. Perspective is not used in engineering except to present an "artist's conception." Because objects appear smaller the farther they are from the viewer, there is no distortion and the drawing "looks right." Perspective drawings are more time-consuming to prepare than other types of drawings. Figure 1.42 is an example of a one-point perspective drawing.

It is referred to as a **one-point perspective drawing** because it has one vanishing point. You can attain more realistic representations if you use more vanishing points. Figure 1.43 shows a two-point perspective drawing.

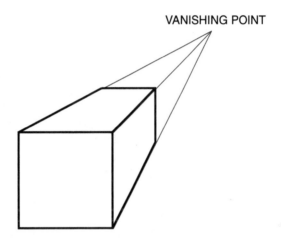

Figure 1.42   One-point perspective drawing

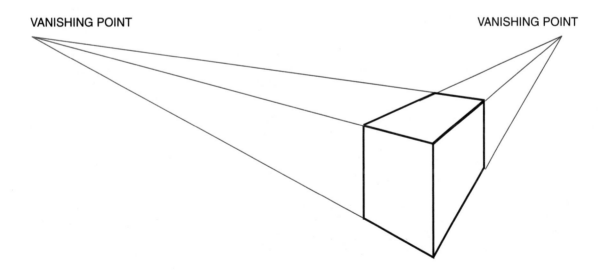

Figure 1.43   Two-point perspective drawing

Three vanishing points could be used, but such a drawing would be complex and time-consuming.

Computer drawing programs can be used to create perspective drawings. Different programs do this in different ways. The mathematical representation of an object in space, seen from different positions, requires a lot of manipulation of data and is best done on a high-powered computer. If the situation warrants the time and effort required to create a perspective, it can be done.

Now that you are familiar with the principles of projection, try the following problems.

Problems

A. Select the best front view and the corresponding top and side views for the objects below. The sides are identified as a, b, c, d, and e. The views are labeled only on object one. Views are the same for all other objects.

Example

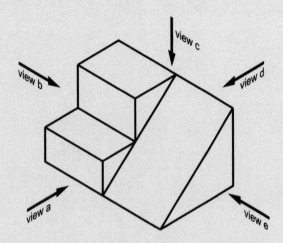

If the front view is **a**, the top view is **c** and the side view is **e.**
If the front view is **e**, the top view is **c** and the side view is **a** (left side).
If the front view is **b**, the top view is **c** and the side view is **a** (right side).
If the front view is **c**, the top view is **a** and the side view is **b** (right side).

Different choices are given here to show how the name of the view depends on the viewpoint. The shape is seen best with front view **e.**

1.

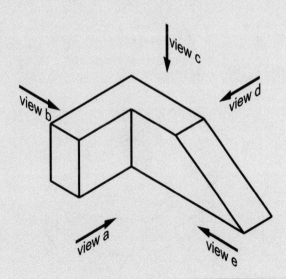

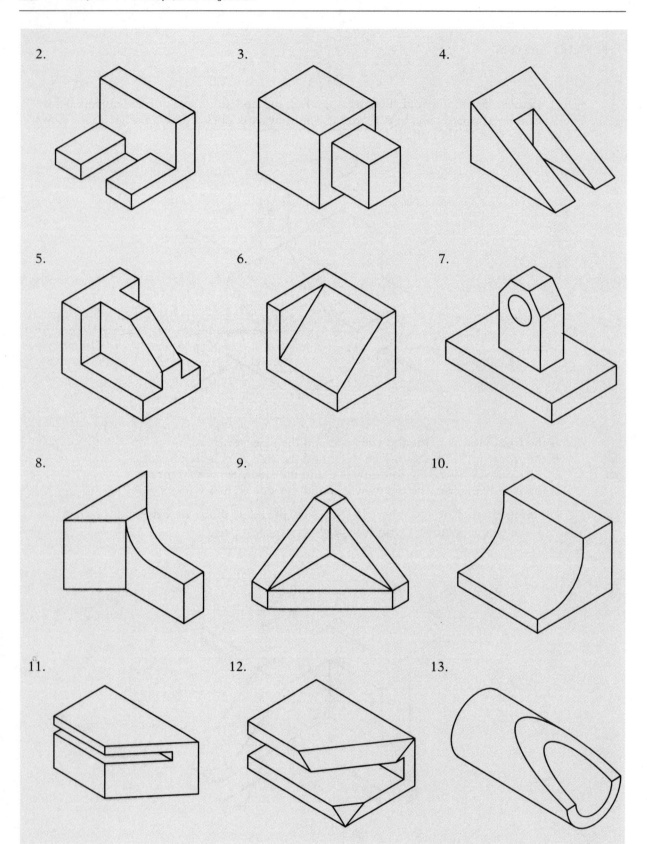

2.

3.

4.

5.

6.

7.

8.

9.

10.

11.

12.

13.

14.

15.

16.

17.

18.

19.

20.

21.

22.

23.

24.

25.

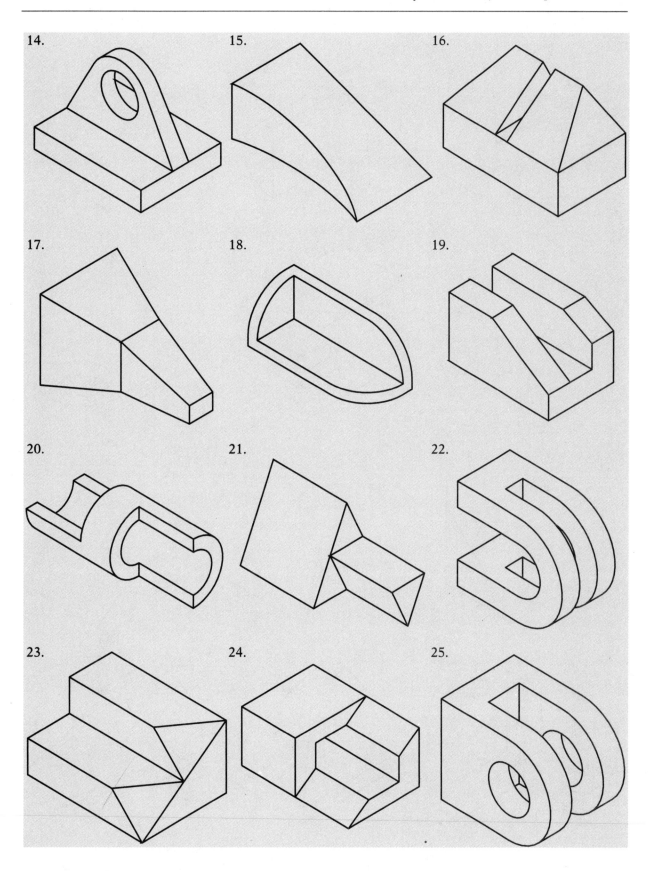

26.

27.

28.

29.

30.

31.

32.

33.

34.

35.

36.

37.

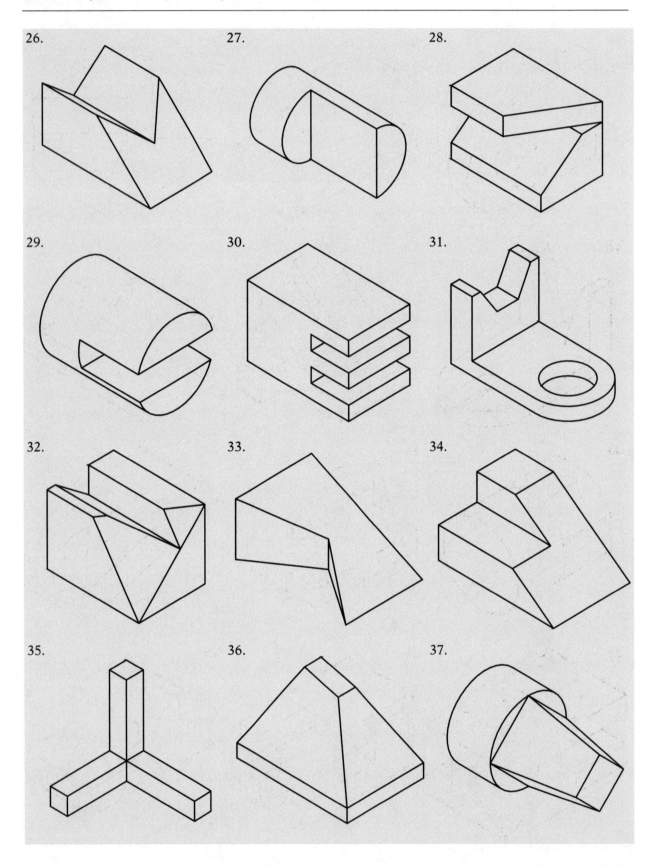

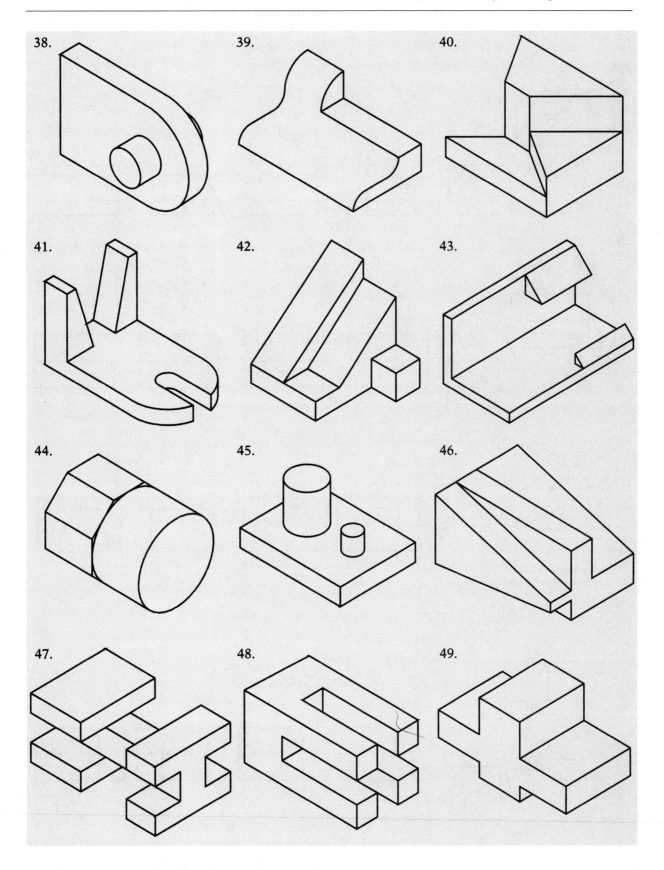

38.

39.

40.

41.

42.

43.

44.

45.

46.

47.

48.

49.

B. Two views of an object are shown. Identify the correct missing view, either a, b, c, or d.

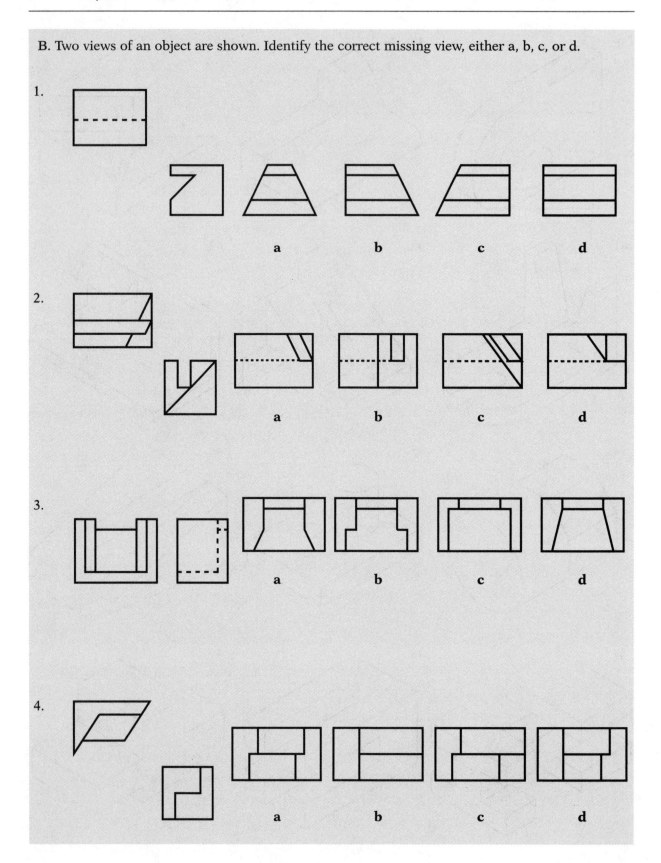

1.

    a   b   c   d

2.

    a   b   c   d

3.

    a   b   c   d

4.

    a   b   c   d

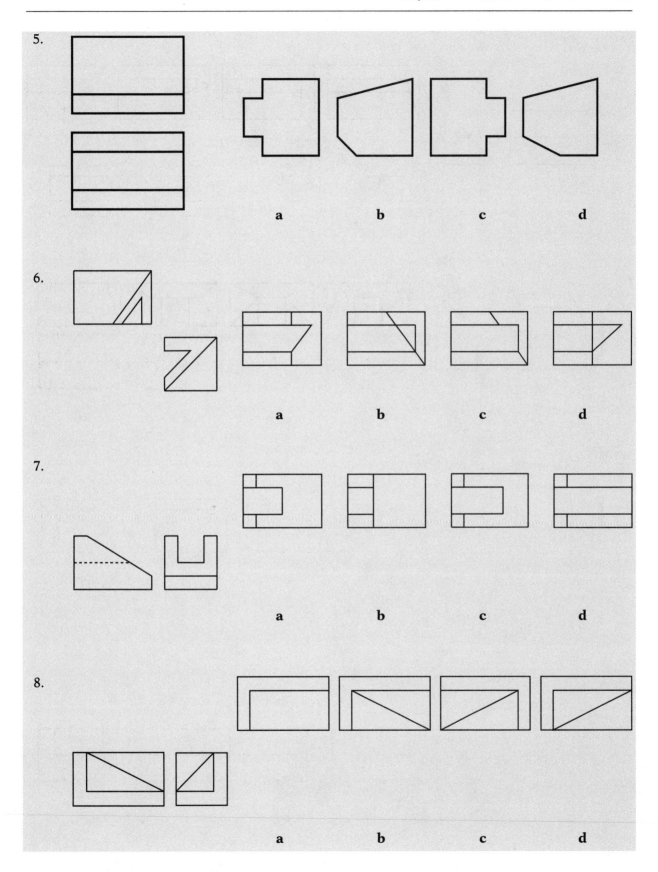

5.

    a   b   c   d

6.

    a   b   c   d

7.

    a   b   c   d

8.

    a   b   c   d

9.

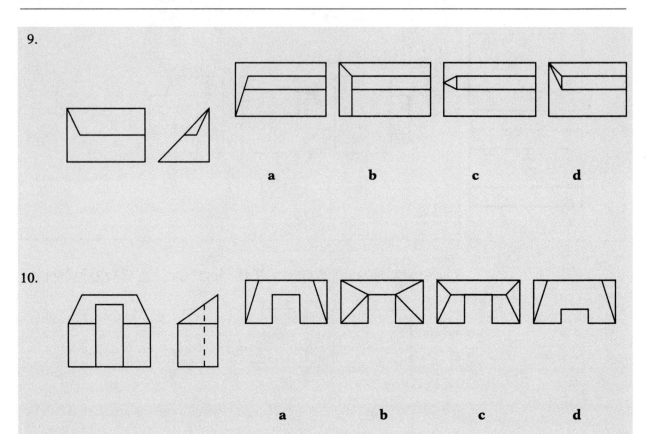

a   b   c   d

10.

a   b   c   d

Chapter 2

Sketching

Using Sketches to Solve a Problem

The first step in solving a problem is to determine what the problem is. Research has shown that successful problem solvers spend more time planning and exploring the problem than they spend actually "doing it." Planning is an important part of the solution to any problem, and sketching is an efficient way of planning.

Engineering problems in real life are not stated as clearly as those in textbooks, where there is usually one correct numerical answer. To determine what the problem is, you must define the shape, dimensions, location and direction of forces, and other information. A sketch can often be the quickest way to see what information is known and what is required.

As an example, consider the problem of showing what shape this equation represents.

$$x^2 + y^2 + Ax + By + C = 0$$

where A, B, and C are constants and

$$D = A^2 + B^2 - C$$

is positive and does not vanish.

There are several ways to approach this problem. You could choose values for A, B, and C, be sure they meet the stated conditions, substitute them into the equation, and determine the value of y for different values of x. You could plot these; however, this would take considerable time.

The equation describes a circle. There are three ways to represent a circle: as a mathematical expression, as a definition, and as a sketch. Which one is best depends on who is looking at it and for what purpose. In this case, a sketch is best because it gives the required information quickly in an easy-to-understand way. Some applications require a mathematical expression, but a sketch of the end result is always very helpful.

Consider the problem of determining the deflection of a roof as a result of a heavy snowfall. Will the snow load cause the roof to deflect more than allowed by local building codes? We will discuss the method of determining this without going into the theory required to solve the problem.

The amount a roof deflects depends on the beams supporting the roof. There is no point analyzing each point along the beam since only the maximum deflection is required. This location can be determined from a **beam diagram**, which shows the length of the beam, loads, and the supports holding up the beam. A fairly simple analysis results in what are called **shear** and **bending moment diagrams**. An example of the analysis of a simply supported beam is shown in Figure 2.1. The diagrams show that maximum deflection occurs at the center of the beam. They also act as a check on the analysis. (The moment diagram is not shown.)

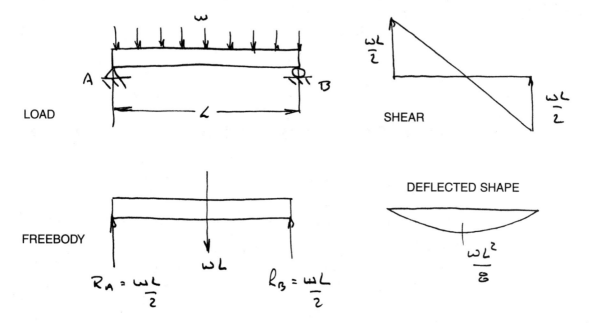

Figure 2.1   Example of beam diagrams

A sketch is the quickest way of putting ideas on paper. The first step in creating any design usually consists of a sketch. Sketches are used to investigate ideas, explain concepts to others, and analyze problems. Preliminary planning can be done quickly with a few sketches. A short time spent on planning can save time and prevent wasted effort. An engineer's time is expensive: to be competitive, an engineering firm must work in the most efficient way possible. For example, a field engineer can sketch design changes in the field and send them quickly to the design office. Analysis to determine the required strength of a component begins with the identification of the forces acting on it. A sketch of a **freebody diagram** to show all forces is the first step in the analysis. Although a field sketch may have to be interpreted at a later time, it is still preferable than relying on memory. The faintest ink is better than the best memory.

The sketch does not have to be an exact representation, but it should show all the necessary information (size, shape, position of components, etc.) so you, or someone else, can understand it at a later time. A pictorial, orthographic, or combination of both can be used.

An example of a sketch used to explore an idea is shown in Figure 2.2. This was drawn when planning an experiment to determine what happens to a domestic water heater during an earthquake. Its purpose was to investigate how the experiment should be set up on the shaker table.

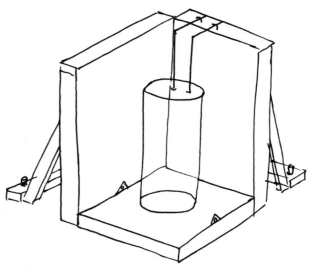

Figure 2.2   Initial sketches to explore solutions to a design problem

Most parts can be formed from simple shapes, such as planes, cylinders, cones, and spheres. Studies have shown that 90 percent of all engineering parts can be formed from these shapes. You need only learn a few techniques to produce a sketch, and all you need is a pencil and paper.

Begin with light construction lines, some of which will be darkened at the last stage. Use a soft pencil, which allows you to draw light and dark lines by varying the pressure on the pencil. You do not need to use different pencils. An HB pencil is a good grade for sketching.

Now let's look at estimating proportions.

Estimating Proportions

A sketch cannot show size precisely, but it should show the features in their correct location and proportion.

The simplest way to estimate proportions is "by eye." For example, in Figure 2.3 is H (height) equal to 50 percent W (width) or 25 percent W?

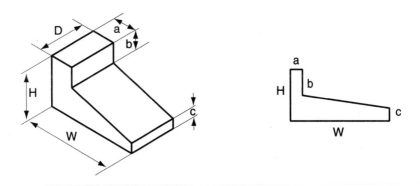

Figure 2.3   Estimating proportions

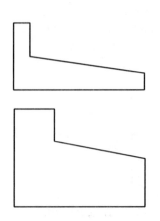

H, W, D (depth), and sides *a*, *b*, and *c* must be shown in the correct proportions. All sides will be compared to the longest side, W.

What size is H relative to W?

H is less than W and greater than 25 percent of W. It is 50 percent of W.

Length *a* is less than 50 percent of W. It is taken as 25 percent of W.

It is easier to compare *b* and *c* with H since they are parallel. Side *b* is taken as 50 percent of H and *c* as 25 percent of H.

These lengths were estimated "by eye" and are not expected to be exact, but the object should be shown with correct proportions. Figure 2.4 shows the same object with incorrect proportions for comparison. Neither representation gives a true picture of the object, although all sides are shown.

Now let's move on to look at sketching straight lines.

Figure 2.4 Distortion due to incorrect proportions

Sketching Straight Lines

A straight line is defined by a starting point and an end point, or any other point if the end point is not known. Locate the starting point and mark it with a cross. Locate an end point on the line. An exact end point can be defined later. Start with the pencil at the starting point and focus your eyes on the end point. Do not watch the pencil. With your eyes focused on the end point, move the pencil toward it. You will probably not hit the end point, but, with practice, you should be reasonably close. Position the paper so that your hand is in a comfortable position. You cannot do a good job if your hand and arm do not move in a comfortable manner.

Example 2.1

Sketch three orthographic views of the object shown in Figure 2.5.

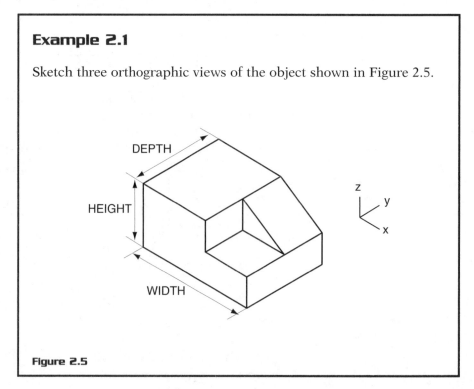

Figure 2.5

1. Decide what elements are important before you show shape and features. An exploded view of the basic shapes combined to create the object is shown in Figure 2.6.

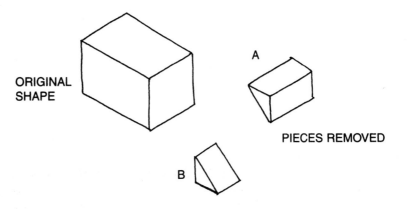

ORIGINAL
SHAPE

A

PIECES REMOVED

B

Figure 2.6 Basic shapes used to form object

There are three basic shapes: a rectangular prism and two triangular prisms. Both triangular prisms, A and B, are "removed" from the rectangular prism. All surfaces are flat.

2. Choose a front view that shows the sloping face in profile.
3. Since no size is specified, there is no point in measuring the isometric (Figure 2.5), so estimate proportions by eye. Front, top, and side views correspond to the x-z, x-y, and y-z planes respectively. Position them as specified in Chapter 1. Use light construction lines, drawn with an HB pencil, to block out sizes and locations. "Remove" prism A from the right. The construction is shown in Figure 2.7.

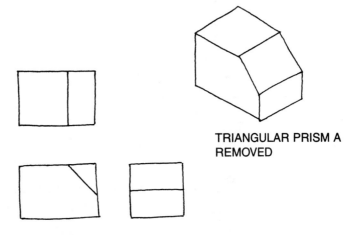

TRIANGULAR PRISM A
REMOVED

Figure 2.7 Sloping face added

4. Remove prism B from the right end and draw lines in all views to represent these surfaces. Views corresponding to this are shown in Figure 2.8.

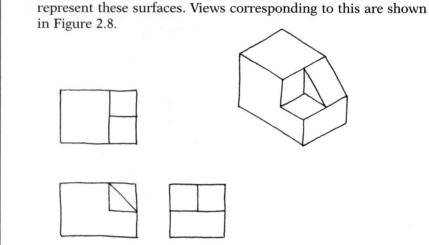

Figure 2.8 All features added as construction lines

5. All lines are shown as construction lines; there is nothing to distinguish them from outlines. The outlines will be darkened, but before this is done, check the sketch to see that it is correct. Does it represent the information given? Are the proportions of the various features correct? Do the features in one view match those in another view? If there are any errors at this stage, correct them before darkening outlines. There is no need to erase construction lines. The finished sketch is shown in Figure 2.9.

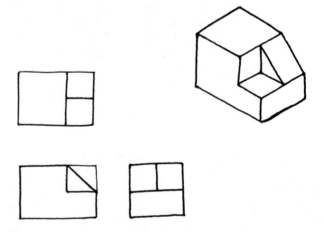

Figure 2.9 Finished sketch with outlines darkened

6. The objectives were to show the shape and the relative sizes of the features. Does this sketch satisfy these objectives? If the answer is "Yes," you are finished. If not, make the appropriate changes so that the objectives are met.

Now let's move on to look at using grids.

Using Grids

Sketching straight lines and estimating proportions are easier when there are lines to follow. Engineering calculation paper often has a non-reproducible grid on one side to aid sketching. Both orthographic and isometric grid paper are available (see Figure 2.10). Grid paper can be easily made using vertical and horizontal lines (or lines at 30° for an isometric grid) produced with a computer drawing program and copied at whatever interval you specify.

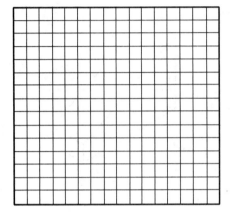

ORTHOGRAPHIC

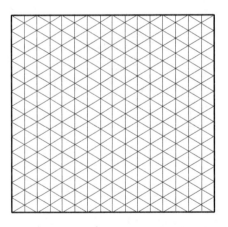

ISOMETRIC

Figure 2.10 Orthographic and isometric grids

Computer drawing programs also have grids to help locate points. An array of dots forms the grid. The user specifies the spacing between dots, which can be changed at any time. Both orthographic and isometric grids can be created. Points can be precisely located with commands that "snap" to a grid point. The grid is not part of the drawing and does not print. Locating points at grid points makes drawing much easier.

Now that you know a little more about using sketches to solve problems, estimating proportions, and sketching straight lines, try the following problems.

Problems

A. Draw or sketch orthographic views of the objects below. Use two or three views to describe the object. Orient the object to give the best front view.

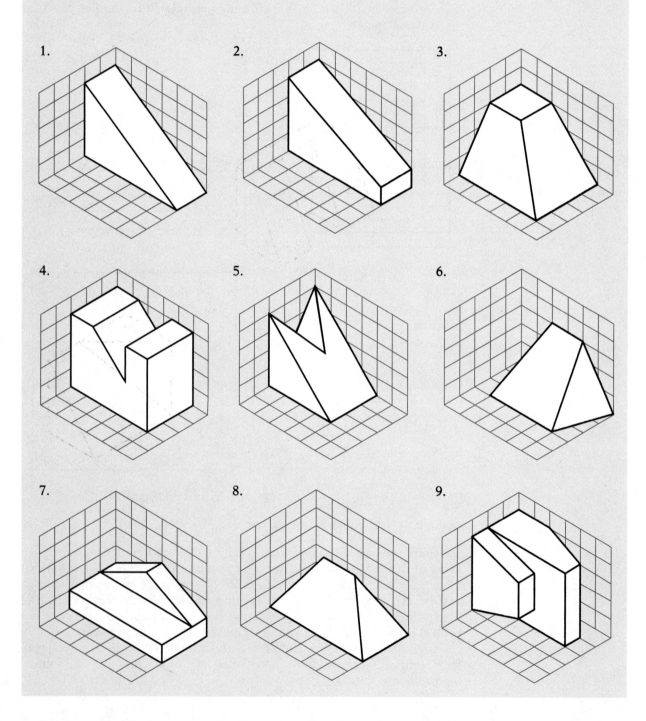

1.

2.

3.

4.

5.

6.

7.

8.

9.

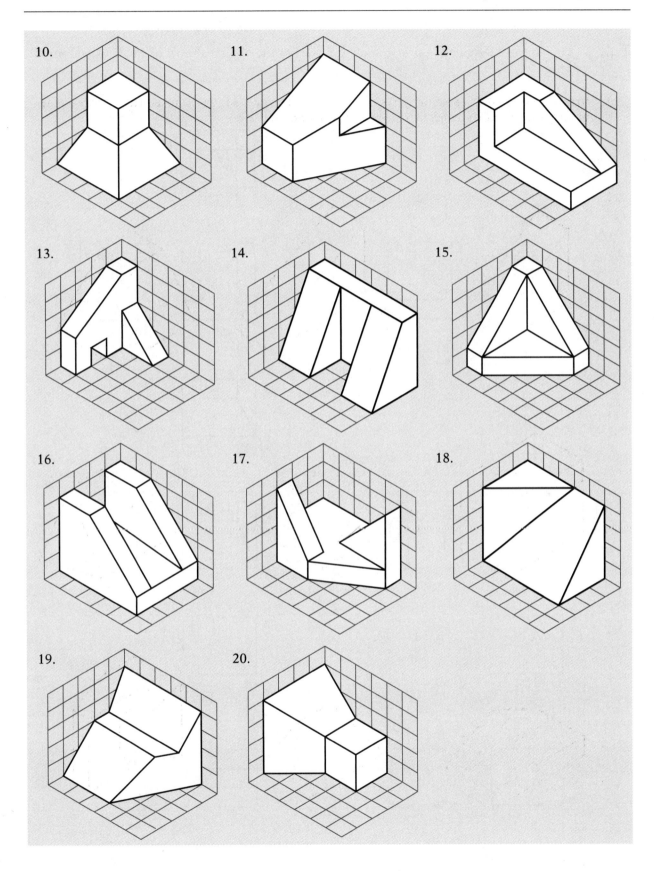

B. Draw isometric sketches, approximately full size, of the objects shown below. The grid may be taken as 5 mm.

1.

2.

3.

4.

5.

6.

7.

8.

9.

10.

11.

12.

13.

14.

15.

16.

17.

18.

19. 20.

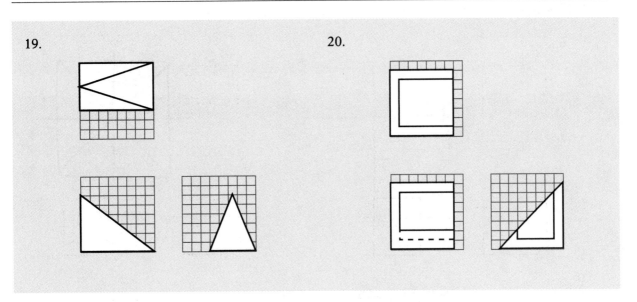

Now let's look at sketching curved lines

Sketching Curved Lines

Since about 90 percent of the shapes found in engineering can be formed by a combination of rectangular prisms, cylinders, and cones, we must be able to sketch a circle or an arc. You already know something about circles: You know the equation describing a circle and you have probably drawn many circles with a compass, a circle template, or a computer drawing program. The problem now is to sketch a circle that looks like a circle. There are techniques to use that will give reasonable results.

A sketch, by its very nature, will not show a perfect circle. It is, after all, drawn freehand. The radius, even if it is known, will not be exact, but no one is going to measure it anyway. The exact size is not important at this stage. A circle is a circle, no matter what size. If you showed your work to strangers and asked them to identify the shape, they would immediately say, "A circle."

We know what a circle looks like and that:

- all circles have a center,
- all circles have a radius,
- any tangent to the circle is 90° to the radius.

Nothing very mysterious here, but can we use this information to construct a circle?

The first step is to locate the center in the desired location. The center is specified by two lines intersecting at 90°. Since these lines identify the center of a circle, they are drawn as centerlines. Size is defined by a square with sides equal to twice the radius, drawn around the center point. Diagonals are added as a check on the size of the square (see Figure 2.11).

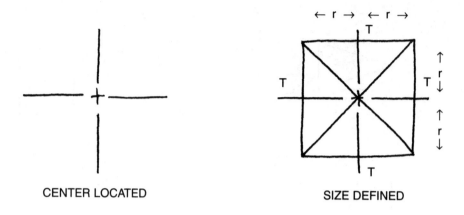

CENTER LOCATED   SIZE DEFINED

Figure 2.11   Initial steps for sketching a circle

The circle will be tangent to the square at the four points identified in Figure 2.11 and the direction of the tangent at these points is known (90° to the centerlines).

All that remains now is to draw the circle.

It is difficult to draw the entire circumference all at once, so draw one quadrant at a time. Position the paper so your drawing hand is comfortable and draw the first quadrant as a light construction line between two tangent points. This line must be an equal distance from the center at all points. You can add more points on the diagonals, if desired, by locating them at a distance "r" from the center. The distance is estimated by eye. Move the pencil between points without touching the paper and, after a little practice, draw a light construction line between the points (Figure 2.12). The direction of the arc must be perpendicular to the centerlines. Repeat this step in the other three quadrants.

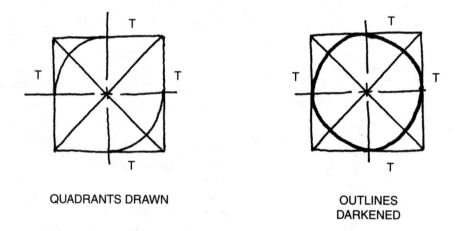

QUADRANTS DRAWN   OUTLINES
                                   DARKENED

Figure 2.12   Final steps in sketching a circle

Darken the outline to distinguish it from construction lines. There is no need to erase construction lines.

You can follow the same process when sketching an arc.

An irregular curve can be drawn by drawing a smooth line through points on the curve. The more points used to define the curve, the easier it is to sketch. Figure 2.13 shows an irregular curve constructed through several points.

Figure 2.13   Irregular curve drawn through points

Example 2.2

The shaft support bracket, shown in Figure 2.14, supports a long shaft which passes through the large hole. The four holes in the base are used to secure the bracket. The bracket will be made of welded steel. A multiview drawing is required for the machine shop. No dimensions will be added.

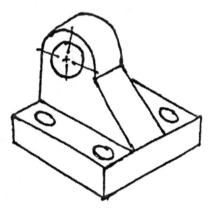

Figure 2.14   Pictorial of object with plane and curved surface

1. Even though the bracket in Figure 2.14 is a fairly simple object, a breakdown of the basic shapes is useful in understanding the problem. Figure 2.15 shows these basic shapes.

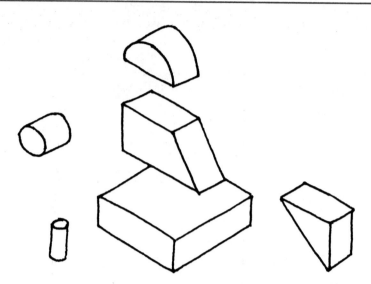

Figure 2.15   Exploded view showing basic shapes

2. Choose the front view so that a right-side view can be used, and the size and the location of holes and other features blocked out. Show centers of circles and arcs by centerlines and define sizes by squares, or, in the case of the arc, by a rectangle. Draw hole centerlines in all views. Figure 2.16 shows the location and size of all holes blocked out.

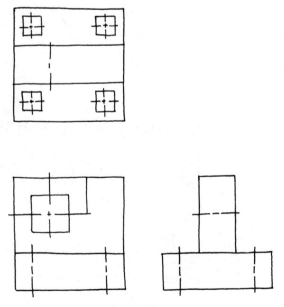

Figure 2.16   Location and size of holes blocked out

3. Add hidden lines as construction lines. If the shapes and sizes are correct, darken the outlines. Figure 2.17 shows the final sketch.

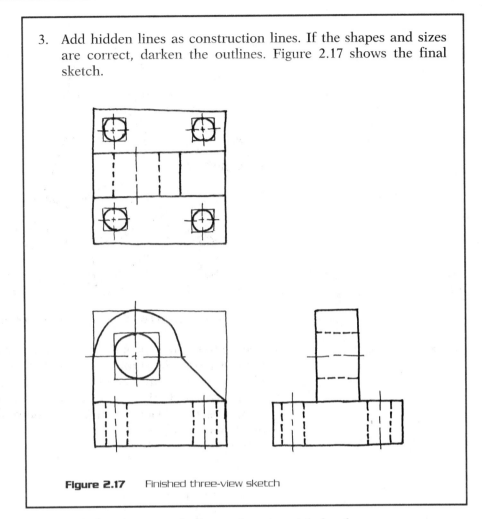

Figure 2.17   Finished three-view sketch

Now let's move on to look at using pictorial sketches.

Using Pictorial Sketches

Pictorials are often used to aid visualization. Some computer programs can create a pictorial from orthographic views, or vice versa, so that the designer can work with whatever is more appropriate. Sometimes it is easier and faster to develop your ideas by sketching a pictorial and then drawing detailed orthographic views. Some computer programs have methods for "sketching." The start and end points of a line do not have to be specified exactly and corners do not have to meet. The program will "clean up" the drawing according to programmed instructions.

Pictorial sketches, like orthographic sketches, are a combination of straight lines, curved lines, and circular arcs. Straight lines can be drawn by locating end points and joining them. Circular arcs require the location of the center and a radius. A circle appears as an ellipse when drawn as an isometric or other type of pictorial.

Example 2.3

The objective is to sketch an isometric pictorial of the object shown in Figure 2.18 so that it can be easily visualized by someone who is not familiar with engineering conventions. Proportions must be correct and the shape must be clearly shown.

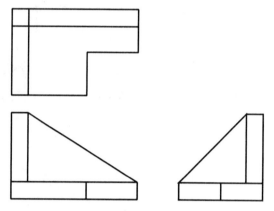

Figure 2.18   Three-view drawing

1. Block out the "space" occupied by the shape, with an enclosing box to define the width, height, and depth. Determine proportions by eye. The height is about two-thirds the width, and height and depth are equal. Block out the "space" with light construction lines as shown in Figure 2.19.

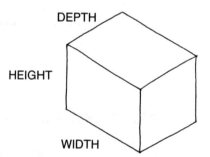

Figure 2.19   Overall size defined by an enclosing box

2. Check the sketch to see that proportions are correct. Does the shape of the box defining overall size correctly indicate the relative lengths of the three sides? Make any corrections before proceeding.
3. The sides of the enclosing box represent the planes of the front (the x-z plane), top (the x-y plane), and side (the y-z plane) views. Transfer the orthographic views to the three sides of the box as shown in Figure 2.20. The box has been "exploded" to show this more clearly.

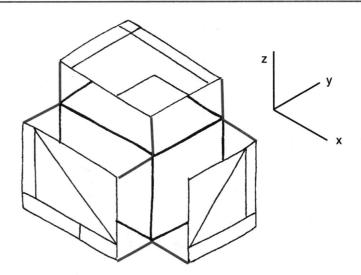

Figure 2.20   Orthographic views drawn on enclosing box

4. Combine the coordinates on the x-y, x-z, and y-z planes to give coordinates (x,y,z) in space. Project points from the projection planes to the object. This is illustrated in Figure 2.21 with points on the base. (Only the base is shown in Figure 2.21.) End points of lines on the base are determined and joined by straight lines.

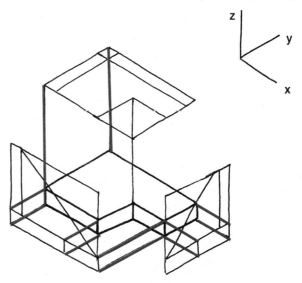

Figure 2.21   Projection from orthographic views to form isometric

5. Draw non-isometric lines (lines not parallel to any axis) by locating the end points and joining them. The completed object is shown in Figure 2.22. Note that there are several groups of parallel lines in this object. Parallel lines always appear parallel, so if they do not appear parallel in any view, it is an indication that an error has been made.

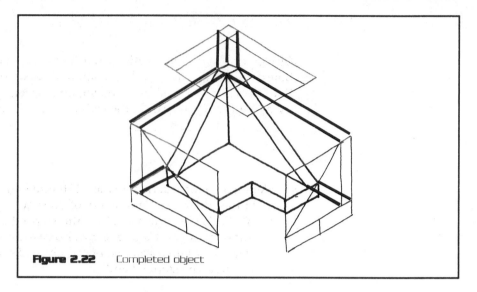

Figure 2.22 Completed object

Now let's move on to look at how to draw circles in isometric.

ORTHOGRAPHIC

Sketching Circles in Isometric

To draw a circle in isometric, follow the same procedure used to sketch an orthographic view, that is, locate tangent points and draw one quadrant at a time. A circle appears as an ellipse in isometric. The enclosing box is drawn in isometric rather than in orthographic. Figure 2.23 shows orthographic and isometric views of the enclosing box. The location of the center can be found by joining the corners and drawing diagonals. The diagonals are not equal in an isometric. The midpoints of each side are joined to locate tangent points.

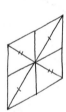

ISOMETRIC

Figure 2.23
Orthographic and isometric drawings of a square

Each quadrant is drawn separately, but the outline is not a constant distance from the center. Drawing the outline as a dashed line can make it easier to sketch the correct shape. The final line is darkened to distinguish it from construction lines. Before darkening the outline, the results should be checked to ensure that the isometric circle looks as it should. Figure 2.24 shows the process and the finished circle.

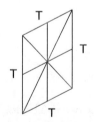

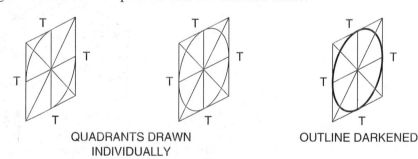

ISOMETRIC ENCLOSURE
DRAWN TANGENTS
IDENTIFIED

QUADRANTS DRAWN
INDIVIDUALLY

OUTLINE DARKENED

Figure 2.24 Steps in creating an isometric of a circle

Now let's move on to look at sketching cylinders.

Sketching Cylinders

Circles are associated with cylindrical components, such as shafts and holes, that are drawn with a combination of straight and curved lines. Cylinders are specified by diameter, not radius, because a diameter is easily measured. (How would the radius of a hole be measured?)

Example 2.4

Sketch a 50 mm diameter shaft, 150 mm long, with one end stepped to 25 mm diameter for a length of 25 mm.

1. Position the object to show the stepped end clearly. Two orientations—a good and a poor choice—are shown in Figure 2.25. The right illustration is a poor choice because it does not clearly show the stepped end.

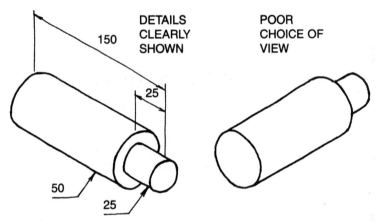

Figure 2.25   Sketch defining sizes and features

2. Block out the space with light construction lines. The construction, a combination of four circles representing the ends of the cylinders, is shown in Figure 2.26.

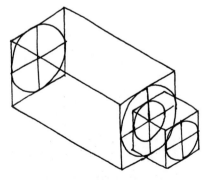

Figure 2.26   Construction of a stepped shaft

3. The finished sketch with hidden lines removed and outlines darkened is shown in Figure 2.27.

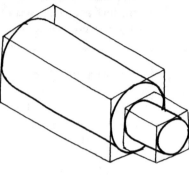

Figure 2.27   Finished sketch

A round hole is also a negative cylinder and is constructed in the same way as a positive cylinder. The difference between a hole and a cylinder is that often the bottom end (or back surface) of the hole is not visible. If the material is thin, a portion of the other end of the hole may be seen on the back surface. In this case, only part of the outline is visible. An enclosing box is drawn on both front and back surfaces, as shown in Figure 2.28. Only the visible portion of the circle on the back surface is darkened in the final step of the construction.

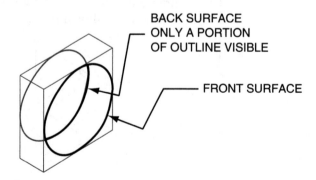

Figure 2.28   Showing both ends of a hole

Now let's move to look at sketching intersecting surfaces.

Sketching Intersecting Surfaces

When two plane surfaces meet at an angle, the line of intersection is a straight line. If one or both surfaces are curved, there is still a line of intersection, but it may not be shown, depending on how the surfaces intersect. If two curved surfaces share a common tangent, the line of intersection is shown in an orthographic view only if the tangent is vertical or horizontal.

If curved surfaces have a common tangent, there will be no abrupt change in the surface and the intersection will not be shown, unless the common tangent is vertical or horizontal. This is shown in Figure 2.29. The line of intersection is not drawn in the pictorial.

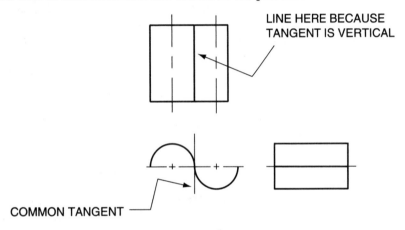

Figure 2.29   Two curved surfaces with common vertical tangent

The common tangent would not be drawn in the front view; it is shown here for illustration only. The line of intersection is seen in the plan view, but not in the side view.

Figure 2.30 shows two curved surfaces with a common tangent that is neither vertical nor horizontal. The line of intersection is not drawn in the pictorial nor in any orthographic view. The common tangent is shown in the front view for illustration only; it would not be drawn in practice.

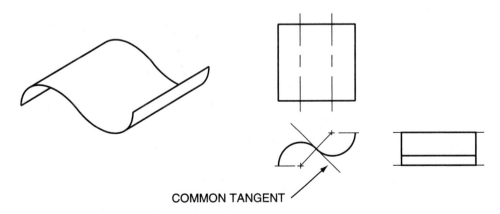

Figure 2.30   Intersection of curved surfaces

When a plane surface is tangent to a curved surface, as shown in Figure 2.31, a line of intersection is not drawn in the isometric or orthographic views.

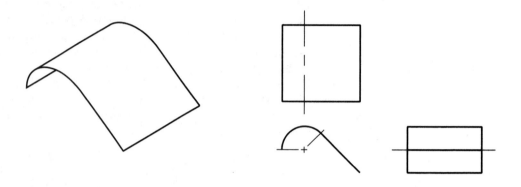

Figure 2.31 Intersection of a plane and a curved surface

Example 2.5

Sketch an isometric pictorial of the object shown in Figure 2.32. Different parts of the object have been labeled for reference.

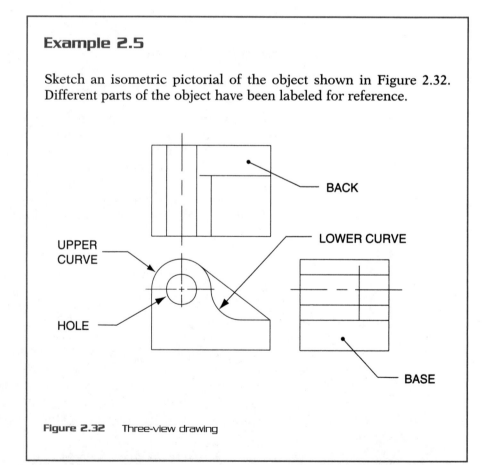

Figure 2.32 Three-view drawing

What is known?

- The basic shapes that make up this solid are:

base	rectangular prism
upper curve	half cylinder
lower curve	quarter cylinder
hole	cylinder
back	triangular prism

- These shapes define overall size, proportions, and features and are used to locate centers, radii, etc.

1. Define overall size by blocking out an isometric frame defining width, height, and depth.

   Take height as 75 percent of width and equal to depth.

   The width of the back is seen in the side view and is taken as one-third of the width.

   These proportions are determined "by eye" and are sufficiently accurate for a sketch. The space occupied by this solid is blocked out using these lengths (see Figure 2.33).

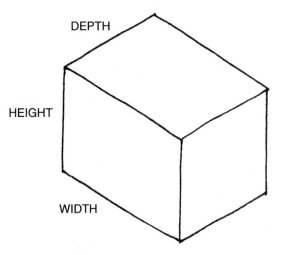

Figure 2.33   Overall size defined

2. Block features into this frame using the basic shapes previously identified.

   The upper and lower curves have a common vertical tangent, indicated by the line of intersection in the plan view.

   The hole and upper curve have a common center.

3. Block the locations of the centers for the circular features and their sizes as shown in Figure 2.34. Only the lower portion of the back is blocked in because only the location of the bottom is known.

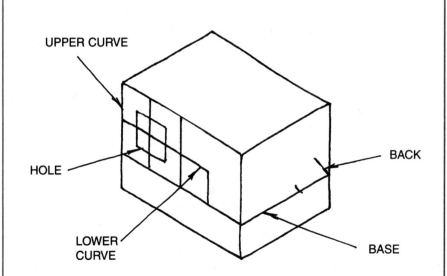

Figure 2.34   Centers and sizes of circular features blocked in

4. Sketch a semicircle, representing the upper curve, and a quarter circle, representing the lower curve, using the method previously described. Darken the outline on the front face to distinguish it from construction lines. The result is shown in Figure 2.35.

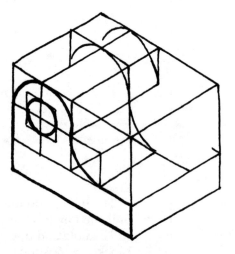

Figure 2.35   Outline defined on front face

5. Project the curved surfaces to the back to complete it. The finished sketch is shown in Figure 2.36, with outlines darkened to differentiate them from construction lines.

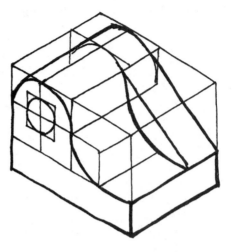

Figure 2.36   Finished sketch

6. An oblique drawing of this object is shown in Figure 2.37. The front view as seen in the three-view drawing is used as the starting point and additions are made to represent depth. There are curved lines that are not on the front surface but these are duplicates of those on the front surface (as with the isometric). They are simply copied in the appropriate location. Because there are curved lines that are not on the front surface, the choice between an isometric and an oblique is not clear cut. It depends on personal preference and the purpose of the sketch.

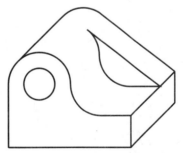

Figure 2.37   Oblique drawing

Now that you know a little more about sketching curved lines, using pictorial sketches, sketching circles in isometric, and sketching cylinders and intersecting surfaces, try the following problems.

Problems

A. Draw or sketch orthographic views of the objects below. Use two or three views to describe the object. Choose the best view for each object.

1.
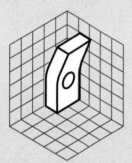

2.

3.

4.
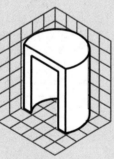

5.
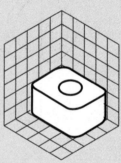

6.
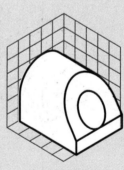

7.

8.
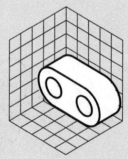

9.

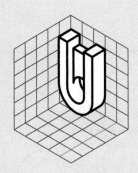

10.

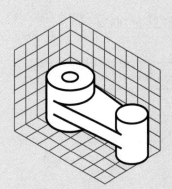

11.

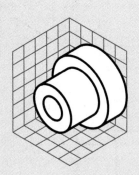

12.

13.

14.

15.

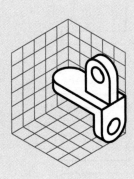

B. Sketch the best isometric view of the following objects.

1.

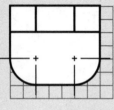

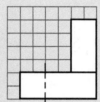

2.

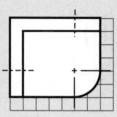

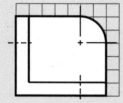

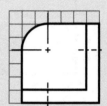

3.

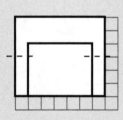

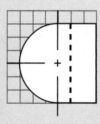

4.

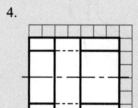

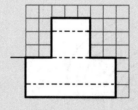

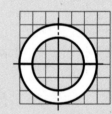

5.

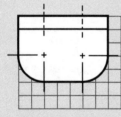

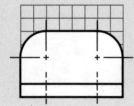

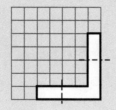

6.

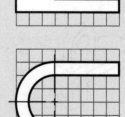

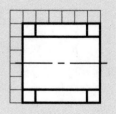

7.

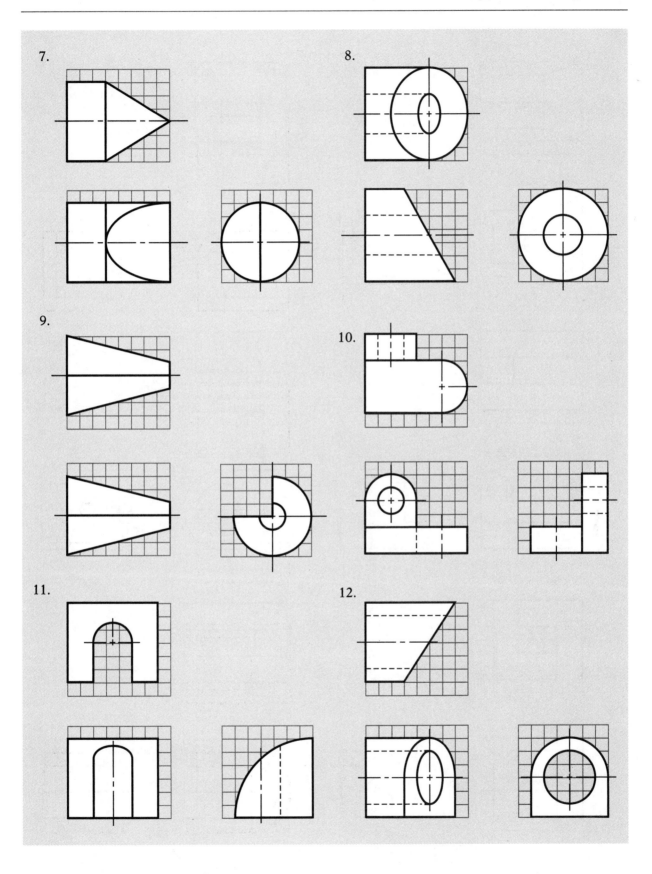

8.

9.

10.

11.

12.

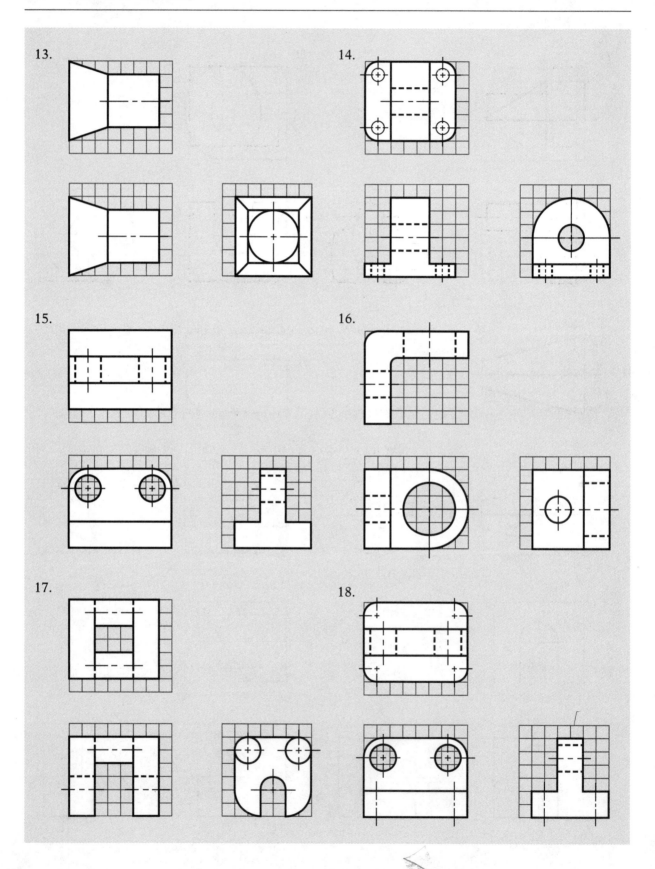

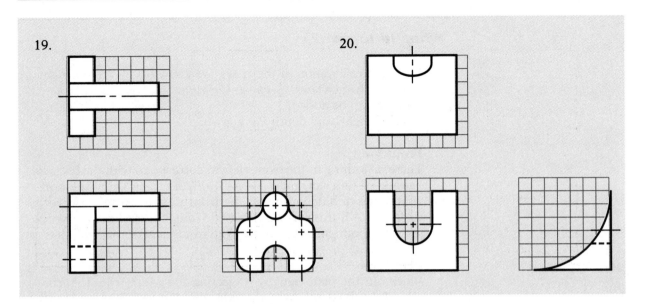

Determining Missing Information

Any missing information must be determined from what is given. With simple objects, you may be able to visualize what is missing, but complicated objects require a systematic approach. Pictorial and orthographic sketches, individually or together, can be used to determine what something looks like.

Example 2.6

Figure 2.38 shows two views of an object. The front view is incomplete and the side view is complete. The front view must be completed and the plan view drawn.

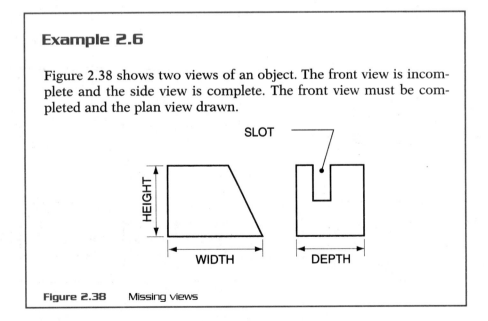

Figure 2.38   Missing views

What is known?

- From the orientation of the views, we know which are the right-side and front views. Because the right-side view is complete, nothing can be added.
- Width, height, and depth are known.

1. **Front View**
   There is nothing in the front view to correspond with the bottom surface of this slot, marked "A" in Figure 2.39. Draw the horizontal line (a hidden line) in the front view to correspond with surface "A." If this surface is not horizontal, more lines are needed, since both ends are visible. The front view is now complete.

2. **Plan View**
   Block out the plan view by projecting the width from the front view. Take the depth from the side view. Figure 2.39 shows work to this point.

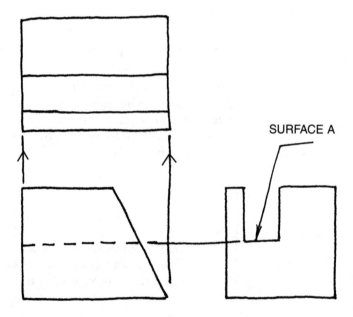

SURFACE A

Figure 2.39   Front view completed and plan view blocked out

3. Find the location of the slot relative to the front surface in the side view (length *s*) and transfer it to the plan view. Locate the slot length and the sloping surface by projecting points *b* and *c* from the front view. Figure 2.40 shows these two steps.

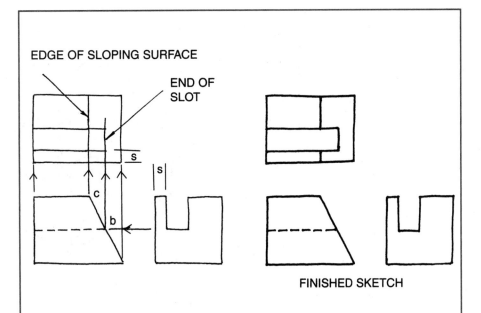

EDGE OF SLOPING SURFACE

END OF SLOT

FINISHED SKETCH

Figure 2.40   Slot and sloping surface located in plan

4. Use an isometric in conjunction with a three-view drawing by sketching orthographic views on the faces of an enclosing box. Figure 2.41 shows the given information added to two faces of an isometric enclosure.

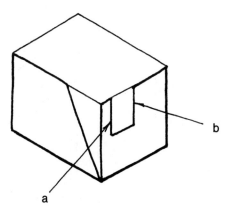

Figure 2.41   Orthographic views transferred to enclosing box

5. Lines identified as *a* and *b* represent the intersection of the vertical sides of the slot with the sloping face. Transfer them from the end to the sloping face. Draw lines representing the slot on the top surface. Transfer the plan view back to the multiview drawing. Figure 2.42 shows the finished isometric sketch.

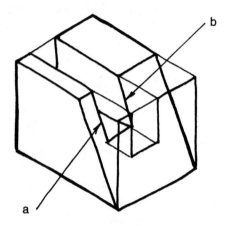

Figure 2.42   Plan view completed

Another approach is to combine basic shapes. Create the object by combining:

- a rectangular prism (a positive volume),
- a smaller rectangular prism (the slot, a negative volume),
- a triangular prism (a negative volume).

Figure 2.43 shows a rectangular prism removed to create the slot.

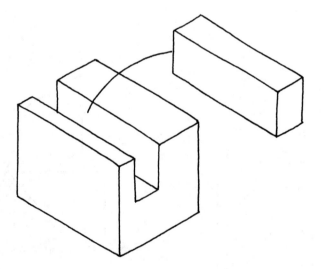

Figure 2.43   Rectangular prism removed to create slot

Remove the triangular prism to form the shape shown in Figure 2.44. Add the line of intersection between this triangular prism and the slot. Show the triangular prism in phantom lines in its original position.

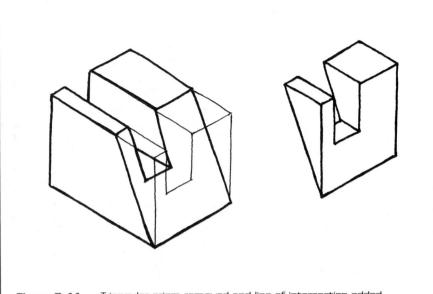

Figure 2.44   Triangular prism removed and line of intersection added

The order in which shapes are removed is not important. The same result is achieved if the triangular prism is removed first.

Now that you know how to determine missing information, try the following problems.

Problems

A. There are lines missing on some of the views shown below; however, there is sufficient information given to determine what the object looks like. Sketch the missing view or the missing lines and an isometric of each object. The objects are shown on a 5 mm grid.

1. Top and side views are complete.
    Complete the front view.

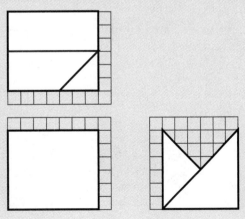

2. Top and left views are complete.
Draw the front view.

3. Top and left views are complete.
Draw the front view.

4. The front and right views are complete
Draw the top view.

5. Top and right views are complete.
Draw the front view.

6. Top and front views are complete.
   Draw the right view.

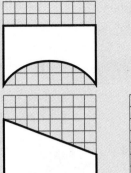

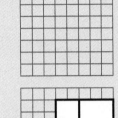

7. The right view is complete.
   Complete the front view and draw
   the top view.

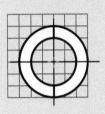

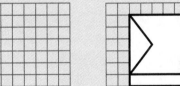

8. The top view is complete.
   Draw the front view and complete
   the right view.

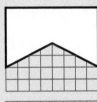

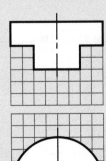

9. The top view is complete.
   Complete the front and side views.

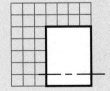

10. The side view is complete.
    Complete the front and top views.

11. The front and side views are complete.
    Complete the top view.

12. Complete all views.

13. Right view is complete.
    Complete the front view and top views.

14. Top and side views are complete.
Draw the front view.

15. Plan and side views are complete.
Draw the front view.

16. Plan view is complete.
Complete side and front views.

17. Front and side views are complete.
Draw the top view.

18. Plan and side views are complete. Draw the front view.

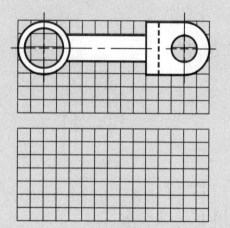

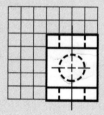

19. Both views are complete. Draw the third view.

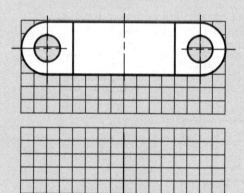

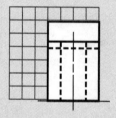

20. Both views are complete. Draw the missing view.

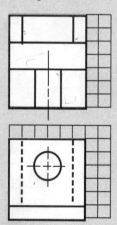

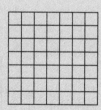

Chapter 3
Sectioning

We have seen how the outside of an object can be shown, but what about internal features? Not all objects are solid, so internal details must somehow be specified. A simple example is shown in Figure 3.1. There is a hole in the block, but no way of knowing anything about the hole. It may only go to the middle of the block, for example. Hidden lines can be used to specify internal features, but they can be confusing, particularly if there are a lot of them.

Sectioning Process

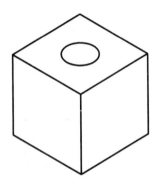

Figure 3.1
Object with internal features

Internal details are shown with a process called **sectioning**. A section of the object is "removed" to show the internal details. Imagine that the block in Figure 3.1 is cut in half through the middle of the hole. Figure 3.2a shows the line on which the block is cut. The front half of the block is removed and the back half is drawn as it would appear after the front half has been removed. Internal features, drawn as solid lines, are now clearly visible (Figure 3.2b).

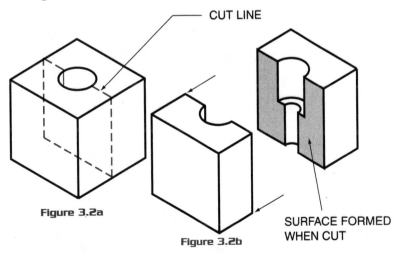

CUT LINE

Figure 3.2a

Figure 3.2b

SURFACE FORMED
WHEN CUT

Figure 3.2   Sectioning to show internal features

To indicate that the front half has been removed and it is not really "half a block," section lines are added to the "surface" formed by the "cut" (see Figure 3.3). Section lines are thin lines, drawn with a straight edge or a computer program. The pattern of the section lines identifies the material. The pattern in Figure 3.3 indicates steel.

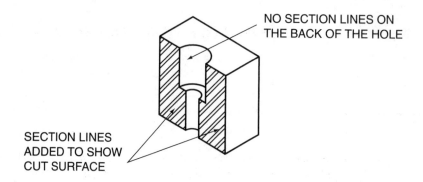

NO SECTION LINES ON THE BACK OF THE HOLE

SECTION LINES ADDED TO SHOW CUT SURFACE

Figure 3.3   Section lines added

Figure 3.4 shows an orthographic view of the block. Internal details are shown by drawing the front view as a section view. The location of the imaginary cut is shown in the plan view by a **cutting plane line**. The arrows at each end of this line show the direction of the eye, and the section is identified as "Section A-A." The front view is drawn as if the front half has been removed and is also identified as Section A-A. If it is obvious where the section has been taken, the cutting plane line can be left out. We will put it in until we become more familiar with sectioning. The cut surface is shown with a section line symbol, in this case, the symbol for steel. Section lines are shown only on the "surface" formed by the cutting plane line and not in the hole. This is called a **full section** because the section is taken through the whole block. Other types of sections will be discussed later.

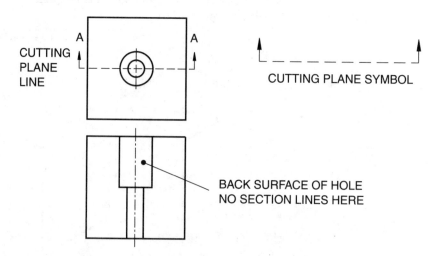

A   A

CUTTING PLANE LINE

CUTTING PLANE SYMBOL

BACK SURFACE OF HOLE NO SECTION LINES HERE

Figure 3.4   Front view as a full section

The methods used for sectioning have been standardized. Understanding the conventions of sectioning make it easier to do and understand. The appropriate standards and conventions must be used so that an engineering drawing can be understood by others.

Let's first of all look at sectioning symbols.

Sectioning Symbols

A unique section line symbol (also called a **hatch pattern** when done with a computer program) can be used to indicate material; however, this is not always done. A simple section line symbol is often used to show that a section has been taken. When section lines are added with a computer program, it is easy to use the correct symbol, but a simple pattern is often used, and the material is specifiied by a note. There are many sectioning symbols—too many to remember. A general section line symbol, which is also the symbol for cast iron (ANSI 31), is shown in Figure 3.5.

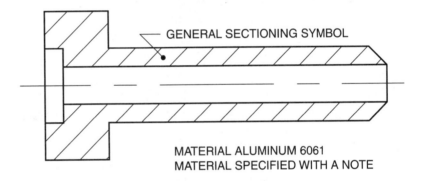

Figure 3.5   Common section line symbol (ANSI 31)

This common symbol can be used if only one part is being sectioned. It has the designation ANSI 31. Sectioning symbols as specified by the ANSI (American National Standards Institute) are given numbers such as ANSI 132 (steel). Some common sectioning symbols and their designations are shown in Appendix B. Computer graphics programs have a library of sectioning symbols, so you do not have to remember all of the symbols and their designations.

Section lines are not added randomly. They are placed so that there is an angle between the section lines and the outlines of the object; however, the angle can vary, depending on the shape of the part being sectioned. The user determines the angle, based on the part being sectioned. Figure 3.6 shows two examples of section lines and the effect of different section line angles. The angle should be selected so that section lines are not parallel, or almost parallel, to any section of the outline. This can sometimes be difficult to do and the end result must be a compromise. The objective is to make the drawing as clear as possible.

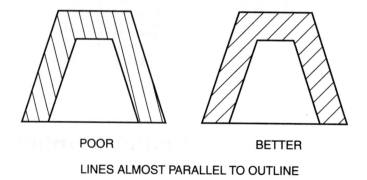

POOR   BETTER

LINES ALMOST PARALLEL TO OUTLINE

Figure 3.6 Determining the section line angle

A common problem when sectioning is to omit lines that are behind the cutting plane line. Lines on the object must be drawn if they are visible. Figure 3.7 shows a front view drawn as a full section to show details of a hole.

BACK OF HOLE

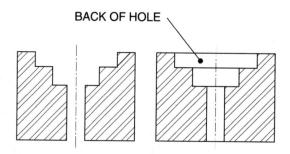

Figure 3.7 Full section showing all visible lines

Lines at the top, bottom, and middle of the hole are on the back surface of the hole. They do not vanish when the block is sectioned and so they must be shown. Figure 3.7 shows what the section would look like if these lines were omitted. There is nothing to connect the two sides of the block. Section lines are not put on the back surface of the hole since this surface is behind the cutting plane line.

Now let's move on to look at locating the section to show the required internal details.

Locating the Section

The location of the section should be chosen to show the required internal details. The previous example used a section through the middle of the hole, which happened to coincide with the center of the block.

Offset Section

Sections do not have to be taken through the middle of the object. They can be taken anywhere. If, for example, internal details of a hole are required, the obvious place for the section is through the center of the hole (see Figure 3.8).

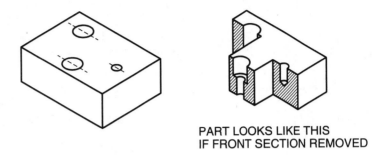

PART LOOKS LIKE THIS
IF FRONT SECTION REMOVED

Figure 3.8   Example of offset sections

This plate has three holes that are not in line. Where should the section be taken to show the shape of each hole? The best location is through the center of each hole, which requires three different cutting planes, as shown in Figure 3.8. Sections should be taken where they will provide the most information, so these are the correct locations; however, three drawings are not required. All sections can be shown on the same drawing as if the holes were in line. Figure 3.9 shows the cutting plane line (in the plan view) where each section is taken. This type of section is called an **offset section**. When this is represented as an orthographic, there is no indication on the section that it is offset. The orthographic representation is shown in Figure 3.9.

Because it is conventional practice to show offset sections in this way, the cutting plane line is sometimes omitted. When in doubt as to whether to show the cutting plane line, put it in.

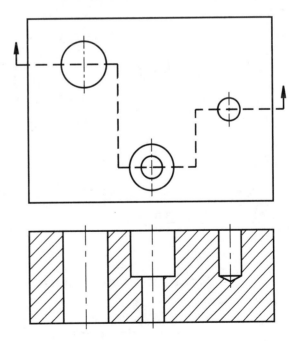

Figure 3.9   Offset section

Half Section

So far, all of the examples shown have been full sections—the section has been taken through the whole object. If there is an axis of symmetry, only one side need be drawn. For obvious reasons, this is called a **half section**. A half section of a simple object is shown in Figure 3.10.

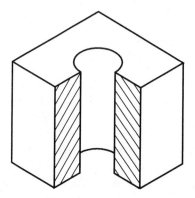

Figure 3.10 Half section

Figure 3.11 shows a cylindrical part with internal features shown as a half section. A 90° segment of the cylinder is "removed" and the front view shows the interior details. Hidden lines on the left side of the hole have been omitted. Since the hole is shown in section, there is no need to show the same features as hidden lines.

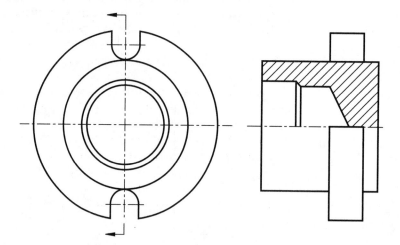

Figure 3.11 Half section

Broken-out Section

Figure 3.12 shows a part with an axial hole and a small radial hole on the top. This could either be shown as a full section or as a half section. But because there are no other interior details, there is no reason to section the whole or even half the part. A section of the region around the holes is all that is required. A piece is "broken out" to show the holes. This is a **broken-out section**. Figure 3.12 shows a plan and front view of the part. There is no need to show a cutting plane line since it is obvious what is being shown and where it is.

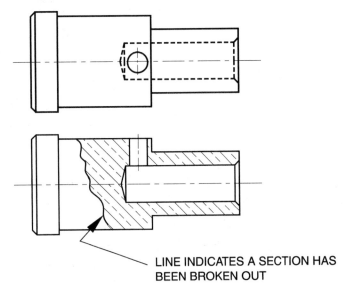

LINE INDICATES A SECTION HAS BEEN BROKEN OUT

Figure 3.12 Broken-out section

Revolved Section

A **revolved section** is the same as a full section except that it is drawn in a different location. Instead of drawing a side view as a section view, the section is positioned on the front view. A revolved section is drawn directly on the view. Figure 3.13 shows a shaft with two cross-sections with revolved sections showing the cross-section. The sections are located on the front view instead of on the two side views.

Only the shape where the section is taken is shown. A conventional side view could also be drawn if required.

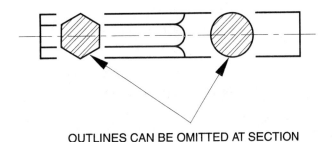

OUTLINES CAN BE OMITTED AT SECTION

Figure 3.13 Revolved section

Removed Section

A section view need not be located on the object itself, as with a revolved section, or in the location of a principal view. It can be located anywhere on the drawing or even on another piece of paper. A **removed section** is simply a section located somewhere other than in a "normal" position. Figure 3.14 shows examples of removed sections. The cutting plane lines show where the section is taken and identify the section (Section A-A, Section B-B). The arrows show the direction of the eye.

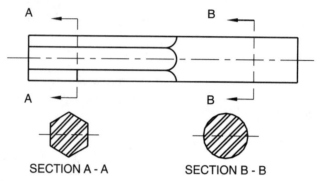

SECTION A - A   SECTION B - B

Figure 3.14   Removed sections

Sections of large items, like buildings, are often drawn on another piece of paper. The cutting plane line indicates where the section is taken. There is usually a note on the drawing with the cutting plane line indicating the drawing on which the section can be found. Figures 3.15 and 3.16 show examples of this.

Now let's move on to look at some standard drawing conventions that can be used in engineering graphics.

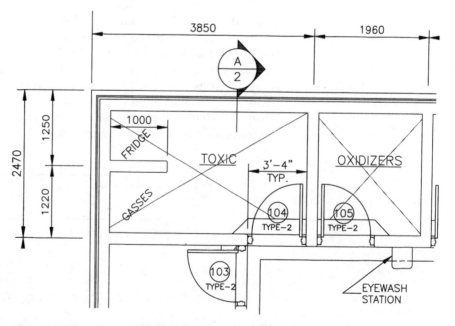

Figure 3.15   Plan view of a building

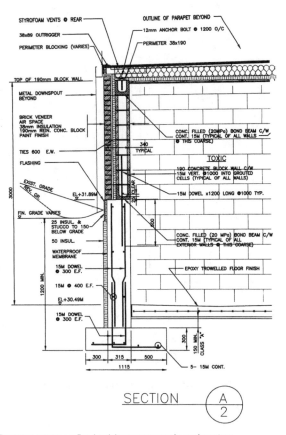

Figure 3.16   Section views of a building on another drawing

Standard Conventions

In addition to providing standardization, conventions can make things easier to do. Some features are not shown as they would actually appear but are simplified to make them easier to draw. Both the person creating and the person reading the drawing must know these conventions for the drawing to make sense.

We will be looking at the standard conventions for conventional breaks, rotations, and thread symbols.

Conventional Breaks

If a part is long (a 1 m round shaft, for example) and there are features at each end to define, only the ends need to be drawn; however, there must be some way of showing that they are the ends of a long shaft. One could draw the whole shaft, but plotting a full-size drawing could present problems with paper size. The solution is to draw the ends and part of the center with a **conventional break**, as shown in Figure 3.17. The shaft is threaded at each end. The symbols at the ends indicate threads (more on threads later). The length of the shaft is specified, but is not drawn full size.

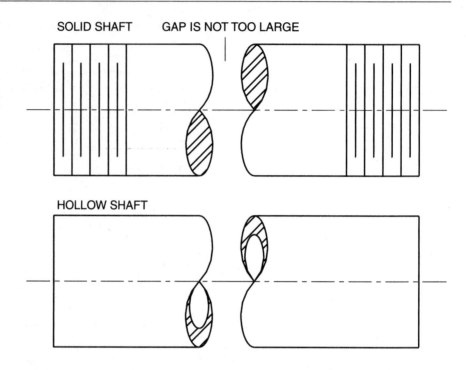

Figure 3.17   Conventional break

The break lines have the shape shown. If a computer is not used, break lines can be drawn freehand. It is not important that the shape be exact as long as it looks like the example shown.

Rotations

Figure 3.18 shows a circular plate with holes spaced at 120° (a common hole spacing). If the side view is drawn following the principles of projection, it is not only confusing to look at, but also time-consuming to draw. So the section is drawn as if the holes were rotated to a location where they show as a true cross-section and the diameter can be seen. The diameter is seen in both plan and section views. Hidden lines are not included.

The same convention is used on other features arranged in a circular array. Figure 3.19 shows a common situation of supports (called **webs**) for a hub. If corresponding points are projected from the front view, the side view is confusing and time-consuming to draw. The side view is therefore simplified by rotating the webs so they appear full size in the side view.

The same applies if the side view is shown as a full section (see Figure 3.20). The webs are drawn as if they were on the cutting plane line. The webs are not sectioned, but there are no internal details to show anyway.

If there are both webs and holes, both are rotated, as shown in Figure 3.21.

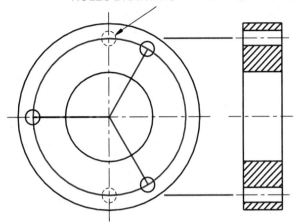

Figure 3.18   Rotation of holes

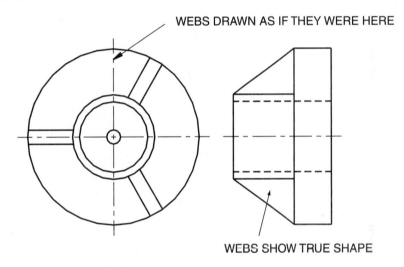

Figure 3.19   Rotation of webs

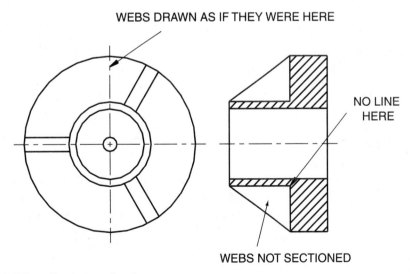

Figure 3.20   Sectioning of webs

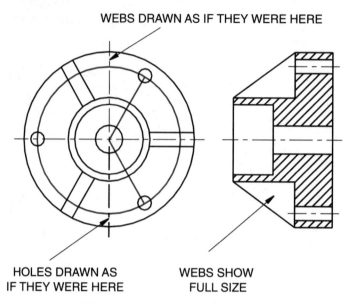

WEBS DRAWN AS IF THEY WERE HERE

HOLES DRAWN AS
IF THEY WERE HERE

WEBS SHOW
FULL SIZE

Figure 3.21 Rotation of holes and webs

The spokes of a wheel present a similar case to webs. Spokes are not sectioned but are rotated so that they are seen in the section. A revolved section could be used to show the cross-section of the spokes. Figure 3.22 shows a drawing of a handwheel with three spokes at 120°.

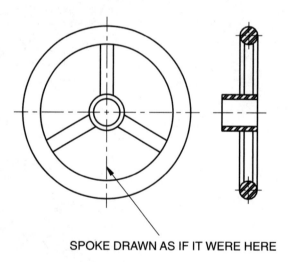

SPOKE DRAWN AS IF IT WERE HERE

Figure 3.22 Rotation of spokes

The full section shows the cross-section of the rim and the hole in the hub as sections. The spokes are shown but not sectioned. Note that two spokes are shown because they are rotated. The cutting plane line has been omitted from the front view since it is obvious where the section is taken. The section view would be the same no matter where the section were taken in this application.

Thread Symbols

For the sake of simplicity and speed, threads are represented by symbols so that each individual thread is not drawn. There are two symbols, schematic and simplified, for use in different applications (see Figure 3.23).

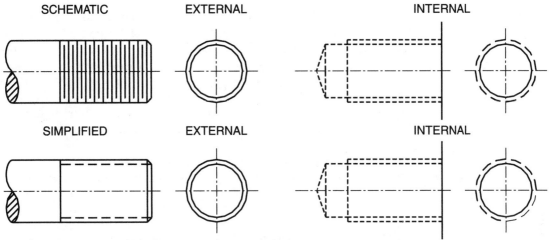

Figure 3.23   Thread symbols

The only information required to show a thread is the size and the length. Size is specified by the maximum diameter. An external 1/2 inch thread has an outside diameter of 0.50 inches. Screw threads will be dealt with in more detail later; at this stage, all we want to do is show how they are represented by symbols.

The lines, which represent the diameter at the base of the thread (called the **root diameter**), are drawn so that they are in the correct proportion for the size of the thread and "look right."

The methods of representing threads in section views are shown in Figure 3.24.

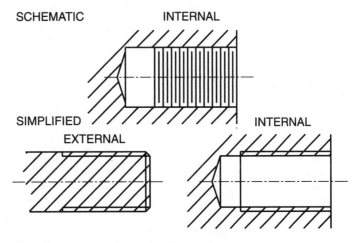

Figure 3.24   Thread symbols in a section view

Now let's move on to look at sectioning in assembly drawings.

Assemblies

All component parts must be drawn in order to make them. An assembly drawing shows how they are put together. In order to save time, only those parts with internal details are sectioned in an assembly drawing. Parts without internal details are standard common items like nuts and bolts. Sectioning these parts serves no purpose. Even if there are internal features, as in the case of a bearing, they are of no concern to the person assembling the parts. Different section line symbols are used to show different materials, and section line angles are varied for clarity.

Figure 3.25 shows the assembly of a shaft passing through the end of an aluminum pipe. The end assembly (steel) is bolted to a flange on the end of the pipe. The steel shaft slides in a bronze bushing held in the end assembly. This drawing shows the relationship of the parts.

The full section shows all of the parts, but not all of them are sectioned. Section lines on the different parts of the assembly are aligned so that they go in different directions on parts that are next to each other. It would be difficult to read if they did not. One part can have section lines in only one direction. The correct section symbols are used because there is more than one material. This means that the areas to be sectioned must be drawn so they can be hatched (when using a computer). The heavy black section between the aluminum and the steel end cap represents a gasket to seal the end. It is too thin for any sectioning symbol, so it is simply filled in. This is the usual way of handling what is called a **thin section.**

Nuts and bolts holding the end cap to the aluminum are not sectioned. These are common standard parts and there is no need to section them. The shaft is not sectioned, except where the conventional break on the left end indicates that the entire shaft is not shown.

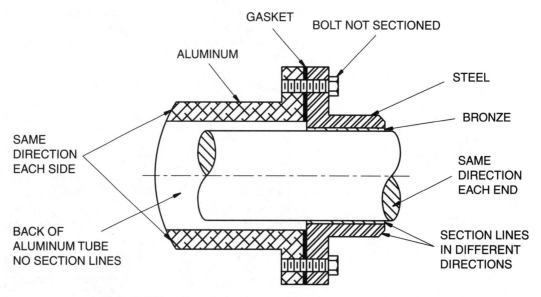

Figure 3.25   Sectioning an assembly

Now that you know the basics of sectioning, try the following problems.

Problems

1. Draw the front view as a full section. Show the cutting plane line in the plan view. Material is steel.

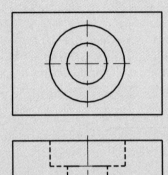

2. Draw a coventional break for the solid shaft.

3. Draw a conventional break for the hollow shaft.

4. Draw the front view as a full section. Show the cutting plane line in the plan view. Material is brass.

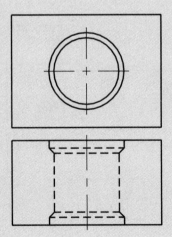

5. Complete the left-side view as a half section. The material is brass.

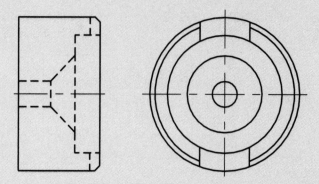

6. The plate has three holes at 120°. Complete the side view as a full section.

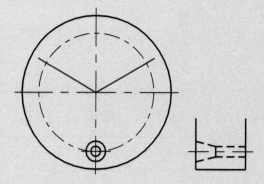

7. Add a broken-out section to show the oil hole in the top of the shaft. Material is steel.

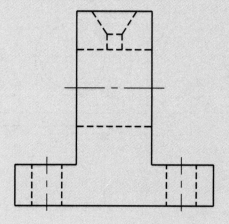

8. Draw the plan view with a broken-out section to show details of the **axial and radial holes.** The radial hole goes half way through the part.

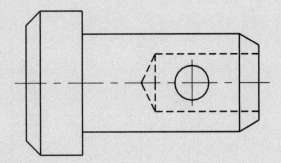

9. Complete the right-side view as a half section. Material is brass.

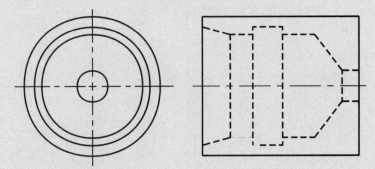

10. Draw the front view as a full section. Show the cutting plane line in the plan view. Material is aluminum.

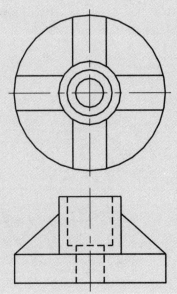

11. Draw the right-side view as a full section. Show the cutting plane line in the front view. Material is steel.

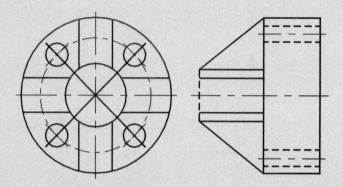

12. Complete the left-side view as a full section. Material is steel.

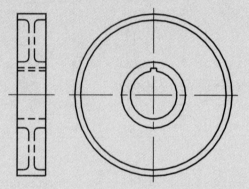

13. Complete the right-side view as a full section. Show the circular cross-section of the spokes with a revolved section.

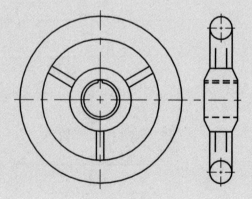

14. Complete the left-side view as a full section.

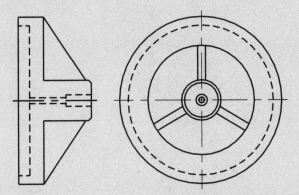

15. Complete the front view as an offset section to show the three holes. Show the cutting plane line in the plan view. Material is aluminum.

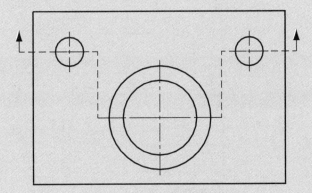

16. Draw the right-side view as a full section showing the internal and external threads.

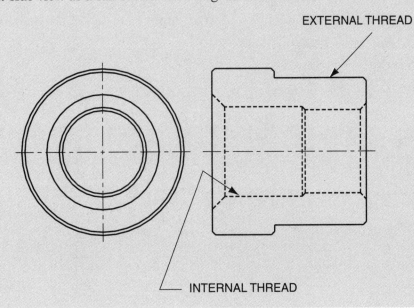

EXTERNAL THREAD

INTERNAL THREAD

17. Draw the right-side view as a full section to show the holes.

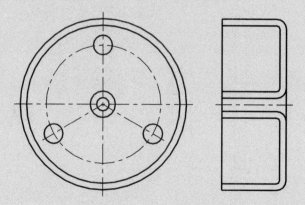

18. Complete the front view as a full section to show the horizontal holes (three).

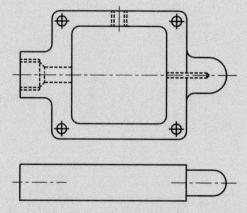

19. Complete the front view as a full section to show the shapes of the three holes.

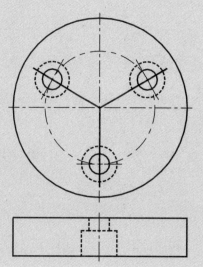

20. Complete the front view with the appropriate sections to show the oil holes in each end and the rectangular cross-section of the center section. Material is steel.

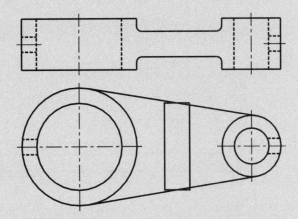

21. Complete the side view as a full section.

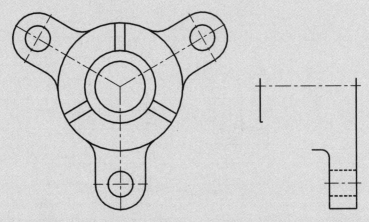

22. Complete the front view as a section view to show the holes. Show the offset cutting plane in the plan view.

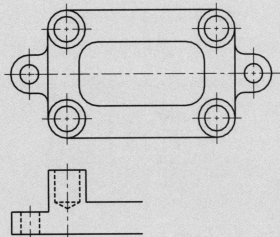

23. Complete the front view as a full section.

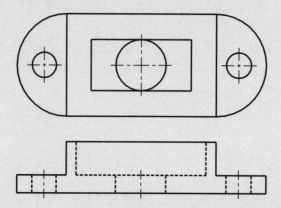

24. Draw section lines on the assembly. The body is steel and the packing nut is brass.

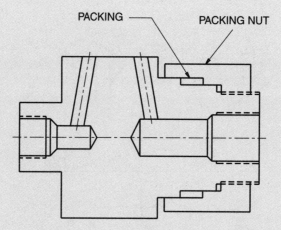

PACKING — PACKING NUT

25. Section the body of the safety valve and the spring. The body is brass and the spring is steel.

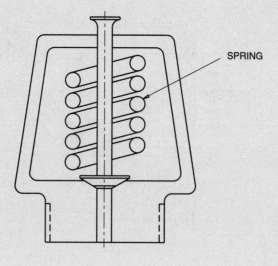

SPRING

26. Complete the front view as a full section. Note that a portion is rotated. Draw revolved sections to show the elliptical shape of the arms.

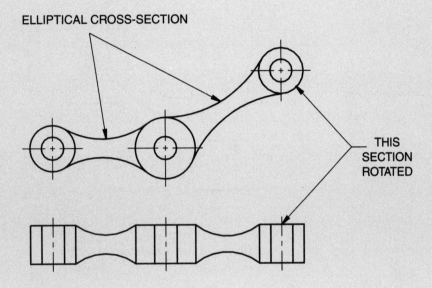

ELLIPTICAL CROSS-SECTION

THIS
SECTION
ROTATED

27. Use broken-out sections in the front view to show the details of the holes. Use a revolved section to show the rectangular center section.

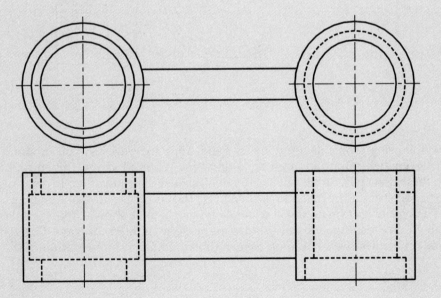

28. Complete the front view as a full section. Show the cutting plane line in the plan view. Material is brass.

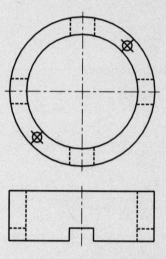

29. Complete the side view as a full section to show details of the hub and rim.

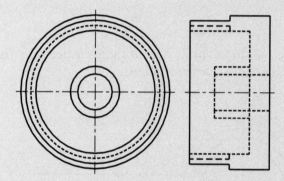

If the following problems are done by hand, copy the pages so you do not have to redraw. There is enough information given to draw the required views. Dimensions may be scaled from these drawings. Some details are not given; use your judgment to determine these, based on what is given. If the material is stated, use the correct section line symbol.

The type of section used will depend on what is to be shown. You must decide the appropriate section technique to use. Some can be done in different ways. Use your judgment to decide on the best way, based on ease of drawing and communication of the information.

30. Use broken-out sections in the front view to show the nylon bushing in each end. The center section is hollow and the walls are 2 mm thick. Use a revolved section to show the cross-section of the center section. Material is steel.

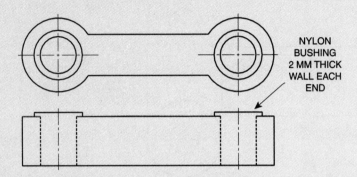

NYLON
BUSHING
2 MM THICK
WALL EACH
END

31. Complete the front view to show details of the hole and the slot.

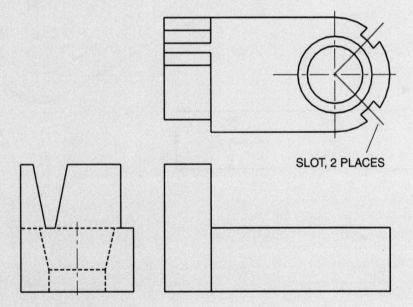

SLOT, 2 PLACES

32. Complete the front view as a full section. Material is steel.

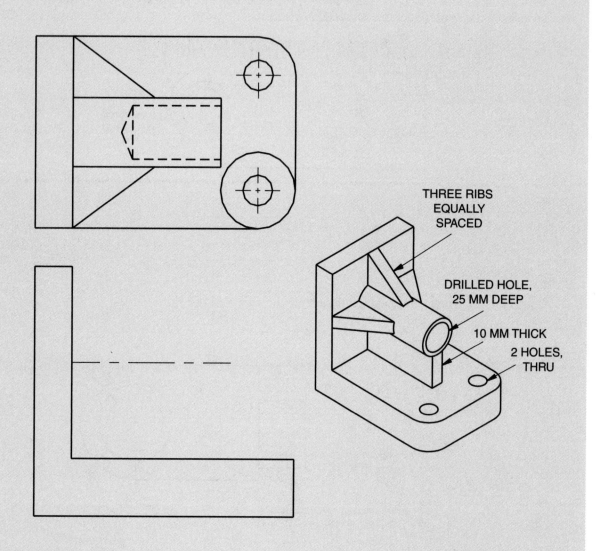

THREE RIBS
EQUALLY
SPACED

DRILLED HOLE,
25 MM DEEP

10 MM THICK

2 HOLES,
THRU

Chapter 4
Dimensioning

Before you can manufacture an object, you need specifications regarding such things as size, the material to be used, and the quantity required.

The size of an object is specified by putting dimensions on the drawing. Two things that must be specified are

- size (length, width, diameter, etc.)
- location (location of the center of a hole or an arc, a corner, etc.).

Size and location are specified by dimensions; other information is covered by written notes on the drawing.

Placing dimensions on engineering drawings and related documents is not a random process. It requires standards.

Standards

Various organizations publish standards that establish uniform methods of dimensioning and tolerancing for engineering and related documents. A **tolerance** is the maximum amount a specific dimension can vary and still be acceptable. The American Society of Mechanical Engineers Standard Dimensioning and Tolerancing, ASME Y14.5M, and the Canadian Standards Association, Dimensioning and Tolerancing, CSA-B78.2, are two examples of standards that apply to engineering drawings. The dimensioning sections of ASME Y14.5M and CSA-B78.2 are essentially the same, but there are minor differences. Other national standards that apply to various aspects of engineering documents are set by ASME, CSA, and other organizations. A list of some of these standards is given in Appendix C.

Computer programs have the capability to place dimensions; however, the user must specify what is to be dimensioned, where the dimension is to be placed, and how the dimensions are to be formatted in accordance with the applicable standard.

In addition to standards, it is important that you understand dimensioning units used on engineering drawings.

Units

Two dimensioning units used on engineering drawings are SI units and US customary units.

SI Units

The common linear unit is the millimetre (spelled millimeter in the USA). The symbol for millimetre, mm, is used only in combination with a number (13 mm) and is always lowercase. The number and symbol are separated by a space.

If all dimensions on a drawing are in either millimetres or inches, you do not need to put the symbol after each dimension. The drawing should contain a note stating **ALL DIMENSIONS IN MILLIMETRES** (or **IN INCHES**).

Rules for using SI units and symbols are given in Appendix I.

US Customary Units

The US customary linear unit used on engineering drawings is the decimal inch.

Besides dimensioning units, you need to be familiar with dimensioning terms.

Dimensioning Terms

Many terms are used in specifying dimensions and tolerances; not all are dealt with here.

Figure 4.1 illustrates some of the terms that you will need.

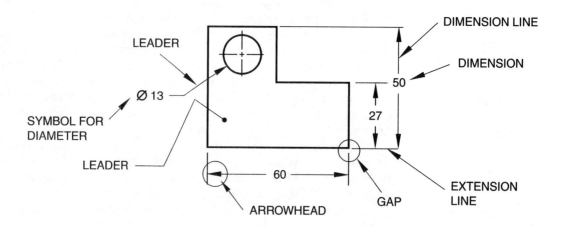

Figure 4.1   Dimensioning terms

Extension lines also called **witness lines** indicate the length. They

- are short, light lines extending from the ends of the feature being dimensioned.
- do not touch the object. There is a small, visible gap (about 2-3 mm) between the object and the extension line.
- extend 2-3 mm beyond the dimension line for the last related dimension.

Extension lines should not cross other extension lines or dimension lines, but this is not always possible. If an extension line must cross another line, there is no break in the extension line.

Extension lines are usually perpendicular to the length being specified, but can be at an oblique angle if there is some reason why they cannot be perpendicular (for example, crowding).

Dimension lines show the extent of the dimension. They are thin lines running parallel to the length being dimensioned. They can be straight (linear dimension) or curved (angular dimension). Dimension lines should not be closer than 10 mm from the object. They should not cross other dimension lines or the outline and should be placed outside of the object (although, again, this is not always possible).

An **arrowhead** is usually placed at each end of a dimension line although other terminators are sometimes used. The length of an arrowhead should equal the height of the numerals used for dimensions, and the width should be about one-third the length. Arrowheads may be filled or open, but the same style should be used throughout the drawing.

Dimensions specify the value of the entity being dimensioned. Numbers should not be smaller than 5 mm. The dimension should preferably be placed in a break in the dimension line, but may be placed above, and parallel to, the dimension line.

Notes give information about other things that must be specified, such as the material and quantity required. Notes are always printed in uppercase letters. To avoid misunderstanding, there are standards for handwritten letters and figures. Notes are usually placed on the right side of the drawing.

Leaders point to a specific feature on the object. If a leader identifies a circular feature, the part of the leader touching the feature must be radial. (It would pass through the center if extended.) The horizontal portion of the leader points to the midpoint of the text or symbols which describe the feature. Leaders pointing to a feature terminate with an arrowhead, and those identifying a surface terminate with a dot. Figure 4.1 illustrates both of these terminations.

Now let's look at linear dimensions and tolerances.

Linear Dimensions and Tolerances

Linear Dimensions

Linear dimensions apply to straight lines or distances and specify the distance between two points. Dimensions can be in any direction. The direction, horizontal or vertical, refers to the paper: a horizontal line is parallel to the bottom of the paper; a vertical line is parallel to the side.

Figure 4.2 shows two methods of linear dimensioning: **chain** and **coordinate dimensions**.

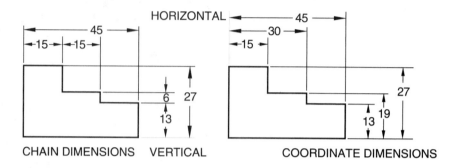

Figure 4.2   Chain and coordinate dimensions

Point-to-point dimensions are called chain dimensions. The starting point for one dimension is the end point of the previous dimension.

Coordinate dimensions are referenced from one point, in effect, assigning coordinates (0, 0) to one point, and specifying the coordinates of other points relative to this origin.

Tolerances

The amount of tolerance acceptable depends on the application. If you were cutting logs, a variation of 100 mm may be acceptable. In other applications, a tolerance of 0.001 mm would be too large.

All measurements have some tolerance, however small. Even the most accurate atomic clocks have a variation of less than one part in 10 billion. The smaller the tolerance, the more it will cost to manufacture and measure. It is advisable to allow the maximum tolerance possible since costs increase as tolerances decrease.

Tolerances are important because parts must fit together. A simple example of how tolerances affect an assembly can be illustrated with a shaft passing through a hole. Figure 4.3 shows a round shaft, with diameter specified as 30 +/- 0.5 mm. This shaft is acceptable if the diameter is between 29.5 mm and 30.5 mm. The tolerance is the maximum variation, in this case 2 x 0.5 = 1 mm. (For illustrative purposes, this is a large tolerance; dimensions are usually given to more than one decimal place.)

This shaft must fit into a hole, which also has a tolerance of 1 mm.

The hole diameter is specified as 30 +/- 0.5 mm. The maximum hole diameter is 30.5 mm and the minimum is 29.5 mm.

The maximum diameter shaft (30.5 mm) will not fit into the minimum diameter hole (29.5 mm), as illustrated in Figure 4.3. This negative clearance is called **interference**. The shaft will not fit into the hole unless it is forced. If this is done it is difficult to take apart. If only one assembly is required, the hole can be enlarged or the shaft diameter reduced so they will fit, but if there are thousands of assemblies this is not practical. The smallest shaft (29.5 mm) will fit into the largest hole (30.5 mm) but the clearance, 1 mm on the diameter, may be too large. With these specifications, some shafts will not fit. The tolerance on the hole and shaft must be changed if every shaft must pass through the hole. Figure 4.4 shows how tolerance on the hole could be changed to allow every shaft to fit.

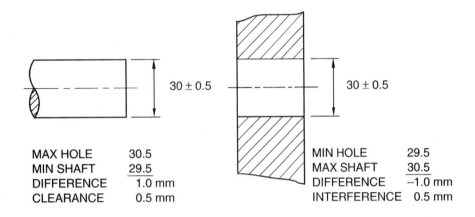

MAX HOLE	30.5		MIN HOLE	29.5
MIN SHAFT	29.5		MAX SHAFT	30.5
DIFFERENCE	1.0 mm		DIFFERENCE	−1.0 mm
CLEARANCE	0.5 mm		INTERFERENCE	0.5 mm

Figure 4.3 Tolerance on shaft and hole diameters

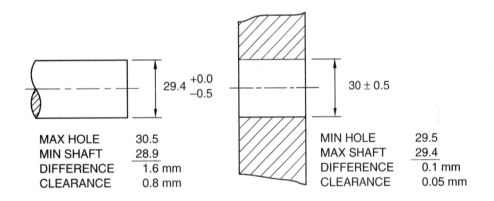

MAX HOLE	30.5		MIN HOLE	29.5
MIN SHAFT	28.9		MAX SHAFT	29.4
DIFFERENCE	1.6 mm		DIFFERENCE	0.1 mm
CLEARANCE	0.8 mm		CLEARANCE	0.05 mm

Figure 4.4 Tolerance adjusted so all shafts will fit

The shaft diameter is specified as 29.4 + 0.0, - 0.5. This is called a **unilateral tolerance** since it can vary in only one "direction." The smallest shaft (28.9 mm) will have a diametral clearance of 1.6 mm in the largest hole (30.5 mm diameter). The maximum diameter shaft (29.4 mm) will fit the smallest hole (29.5 mm) with a diametral clearance of 0.1 mm.

Not only are there tolerances on size dimensions, but there are also tolerances on location dimensions. If tolerances "add up," and sometimes they will, some parts will not fit together and others may be too loose. Figure 4.5 shows how tolerances can combine to affect length and position. Only the variation in each length is shown in this figure. The dimension of each length is given starting at the left side and proceeding to the right.

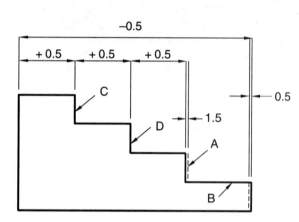

Figure 4.5   Addition of tolerances with chain dimensions

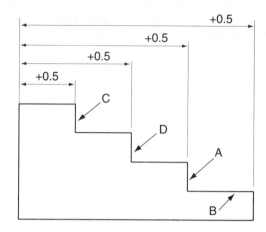

Figure 4.6   Tolerances with coordinate dimensions

If each length has a tolerance of 1 mm (+/– 0.5 mm), edge A could be 1.5 mm (0.5 + 0.5 + 0.5) to the right and B could be shortened by 2 mm (1.5 mm on the left side and 0.5 mm on the right side).

A different method of dimensioning, called **coordinate dimensioning**, can be used to reduce the effect of the addition of tolerances. The same object dimensioned with coordinate dimensions is shown in Figure 4.6.

All dimensions are taken from the same datum, the left side, so that the tolerance on one dimension does not affect other dimensions. Vertical edges C and D will be only 0.5 mm out. When tolerances on size are combined with those for location, the problem can become quite complex.

The tolerances illustrated here are simple tolerances on size and location. Another method of expressing tolerance, called **geometric tolerance**, takes into account the shape of the object. For example, how flat is flat, how round is round, or how straight is straight?

A complete discussion of tolerances and their use is beyond the scope of this book. Ninety percent of specification ASME Y14.5M-1994, deals with tolerances and how to use them. Those who specialize in manufacturing must know more about tolerancing and will see it again when dealing with different machining and manufacturing methods. We must all recognize that any quantity or measurement has some tolerance and as tolerances become smaller, costs increase.

Now let's move on to look at the rules for dimensioning.

Rules for Dimensioning

Because dimensioning must be done in a uniform way, rules and conventions have been established to ensure that everyone "uses the same language." The rules given here are based on ASME Y14.5M and CSA-B78.2 standards.

1. Dimensioning must be complete. There must be no information missing. The user must not be required to make any assumptions or mea-

sure anything directly on the drawing. (This is called **scaling the drawing**.) There must be only one possible interpretation.

Because information is missing from the left-hand drawing in Figure 4.7, it is possible to interpret the dimensions in two different ways.

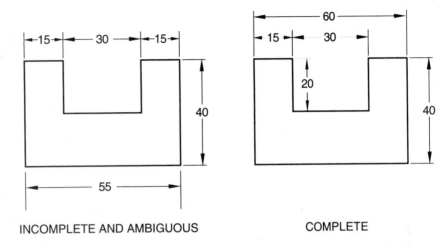

INCOMPLETE AND AMBIGUOUS   COMPLETE

Figure 4.7   Incomplete dimensions and two interpretations

In the left-hand drawing, the width of the cutout is given (30 mm) but the depth is not. The dimensions across the top give an overall width of 60 mm, but the width stated at the bottom is 55 mm. This is ambiguous. Notice how much clearer the right-hand drawing is.

2. Do not add extra dimensions. You cannot dimension everything and "hope for the best." In Figure 4.8, the dimension on the sloping side of the left-hand drawing (58.3 mm) is not necessary since the information required is given by the vertical (30 mm) and horizontal (50 mm) dimensions. (In some situations, extra dimensions are helpful: This will be dealt with later.)

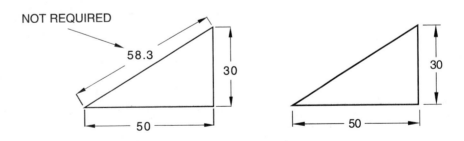

Figure 4.8   Excess dimensions

3. Show dimensions on the true profile view and refer to visible outlines. Do not dimension to hidden lines. There are ways to avoid dimensioning to hidden lines: one is to locate the dimension on another view; another is to use a section view.

The only length that can be seen clearly in the side view (Figure 4.9) is the thickness (12 mm). The size of the hole (20 mm) should not be given on the side view because it is dimensioned to a hidden line. The distance from the bottom to the centerline of the hole (20 mm)—or is it the distance to the corner?—should be given where the shape is seen in profile, in the front view (see Figure 4.10). The location of the hole centerline should be given where the hole is seen as a circle. The height (40 mm) can be seen in the side view, but there is no way of telling what the shape is in this view. The shape of the part is not seen in profile in this view. Height should be shown on the front view.

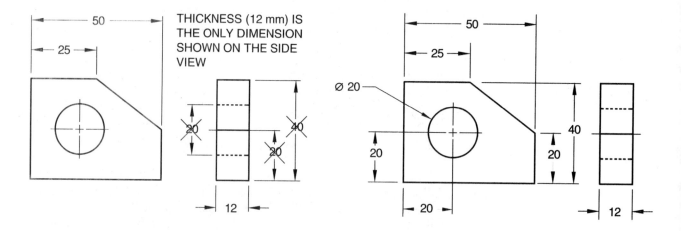

Figure 4.9 Dimensions on the wrong view

Figure 4.10 Dimensions on the correct view

4. Dimensions should be arranged for maximum readability. Figure 4.11 shows the same item dimensioned in two ways. All information is given on both drawings, but it is disorganized on the left-hand drawing.

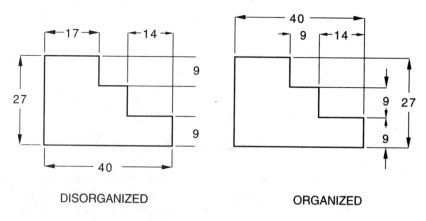

DISORGANIZED

ORGANIZED

Figure 4.11 Disorganized dimensions

Notice that the disorganized dimensions are placed randomly. Some horizontal dimensions are on the top of the part, others on the bottom. When dimensions are organized, as in the right-hand drawing, they are easier to read and understand. Horizontal dimensions can be grouped together at either the top or the bottom of the part. The vertical dimensions should be grouped on the right side. That is where the features (the "steps") are. If there were a right-side view, the dimensions would be between the front and side views. Vertical dimensions are usually located between the front and side views so they can easily be applied to both views.

5. Parts should be described without specifying the method of manufacture. The diameter of a hole would be given without specifying how it is to be made. (There are a number of ways to make a cylindrical hole.) In a situation where a hole must be made in a specific way, the way should be specified.

6. There should be no redundant dimensions, but it is often helpful to have extra information which can be added if identified as extra dimensions. For example, it is often useful to know the overall size. Extra dimensions should be identified as **reference dimensions** by enclosing them in brackets. Reference dimensions are secondary to other dimensions; if there is disagreement, the reference dimension is ignored. Figure 4.12 shows how reference dimensions are identified. Another way of identifying a reference dimension is to add the letters REF after the number (17 REF). The use of reference dimensions should be kept to a minimum.

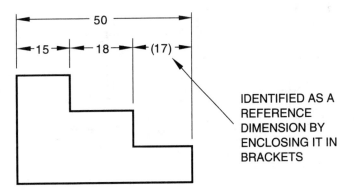

Figure 4.12   Reference dimensions

There are other rules for using dimensions, but those outlined above give a good guide to the intent of dimensions.

Now that you know the basics of dimensioning standards and rules, try the following problems.

Problems

These problems cover size dimensions. They may be done as sketches or as finished drawings on a computer or by any other means.

Dimension the following objects in accordance with the guidelines given in this chapter. The objects are drawn on a square grid. The size can be specified by the instructor or student.

The original sketch or drawing can be passed to another member of your work group (a "checker") for checking. The checker marks all errors and omissions and returns it to the originator for correction.

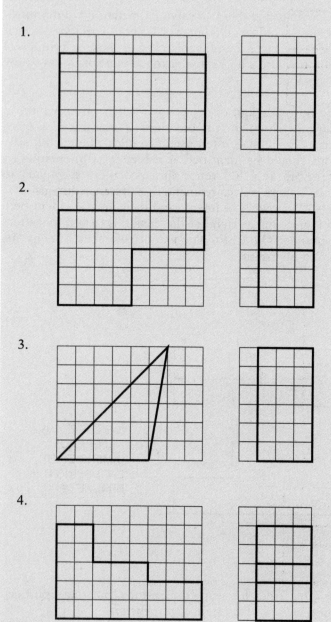

1.

2.

3.

4.

5.

6.

7.

8.

9.

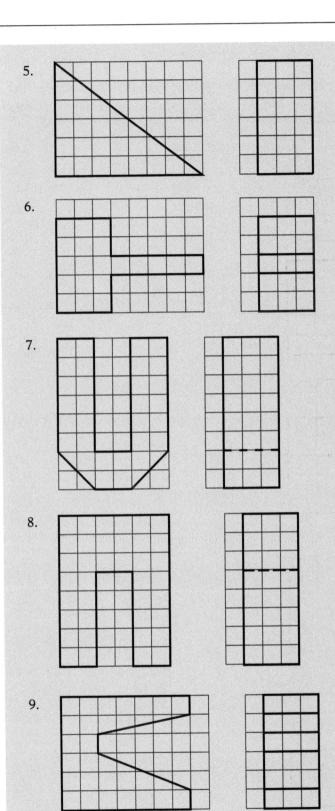

10.

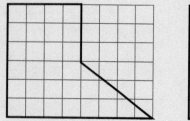

11.

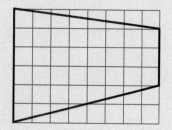

12.

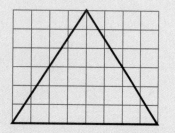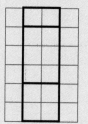

13.

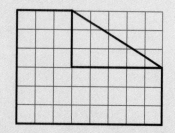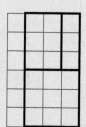

14.

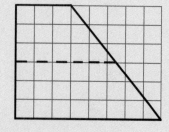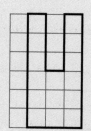

15.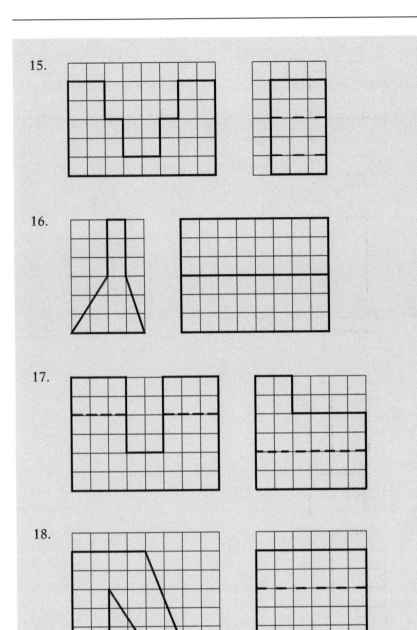

16.

17.

18.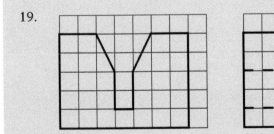

19.

20.

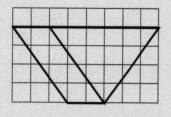

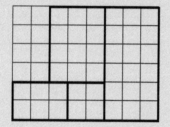

21.

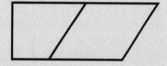

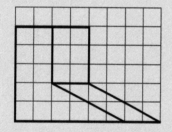
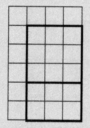

22.

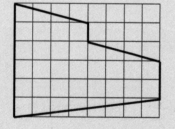
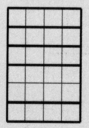

23.

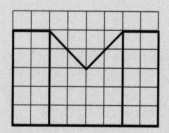
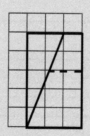

24.

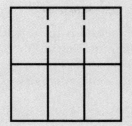

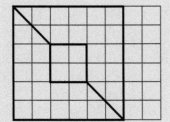

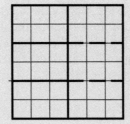

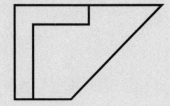

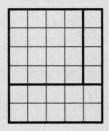

25.

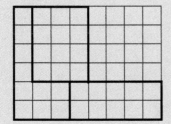

Dimensioning Features

There are accepted methods of specifying dimensions for different features. Using these methods enables you to communicate with other engineers and technical personnel.

In this section we will look at angular and circular dimensions.

Angular Dimensions

Angular dimensions specify the angle between two points. The dimension line is an arc with the center at the apex of the angle and terminating with arrowheads at extension lines (extensions of the two sides of the angle). Angles can be specified in degrees-minutes-seconds or in decimal degrees, as shown in Figure 4.13.

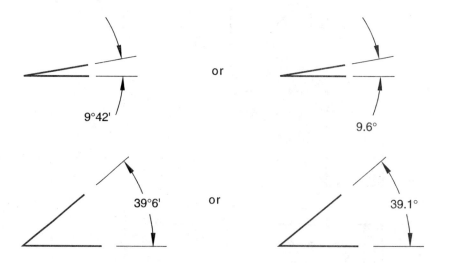

Figure 4.13   Dimensioning angles

Circular Dimensions

Circular features are defined by specifying the location of the center and either a radius or diameter. A center and radius are used for an arc, and a center and diameter for a circle. There are several ways to specify this information, depending on the size of the circular feature and whether it is a solid cylinder or a hole.

Diameter

A solid cylinder is dimensioned where both dimensions, diameter and length, are in the same view as visible outlines. A hole (a negative cylinder) is dimensioned where the circular shape is seen.

Figure 4.14 shows a pulley with three diameters and a hole for mounting on a drive shaft. The diameter and length of each section are shown in the front view. The diameter of the hole is given on the side view where the circular shape is seen. The hole is assumed to go all the way through if there is nothing to indicate the contrary. A note "THRU" is added after the diameter to show that the hole goes all the way through the pulley. The hidden line in the front view also shows that the hole goes through the pulley.

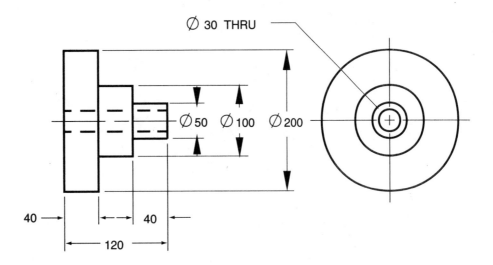

Figure 4.14   Dimensioning cylinders and holes

If a hole does not go all the way through the part (called a **blind hole**), the depth must be specified as shown in Figure 4.15. The depth is the clean depth and does not include the conical portion at the bottom. This conical section results from the shape of the drill used to make the hole.

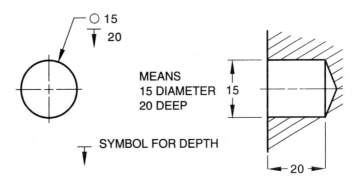

Figure 4.15   Dimensioning the depth of a blind hole

Small holes are dimensioned by a note with a leader pointing to the hole, as in Figure 4.15. The leader is radial and, if extended, would pass through the center of the circle.

Large diameter holes are dimensioned as shown in Figure 4.16.

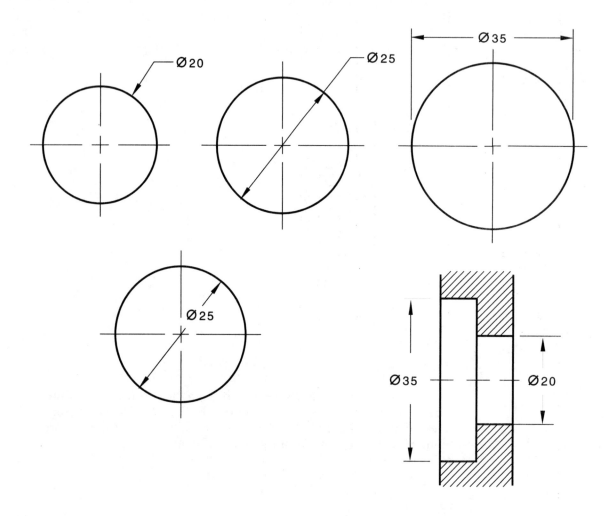

Figure 4.16 Dimensioning large holes

Radius

If a circular feature is not a complete circle, it is specified by the location of the center, the starting point, the end point, and the radius. End points are often defined by other features.

To draw an arc, you must know its center, but the location of the center may not be specified on the drawing. Other information, such as tangent points, must be given to locate the center. This information is usually not explicit, but is obvious from the drawing. If for some reason it is not obvious, the location must be given so it can be located correctly. Figure 4.17 shows different ways to specify a radius without giving a center.

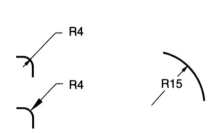

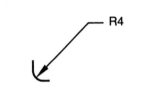

Figure 4.17a **OUTSIDE CORNER**

Figure 4.17b **INSIDE CORNER**

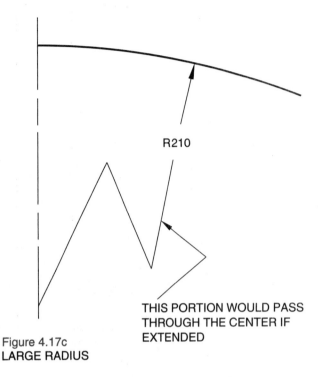

R210

THIS PORTION WOULD PASS
THROUGH THE CENTER IF
EXTENDED

Figure 4.17c
LARGE RADIUS

Figure 4.17 Dimensioning a radius without the center

The radius is given as a dimension with a line from the center, even though the center is not marked. If the arc is not large enough to include the dimension, a leader is used either "outside" or "inside" the arc, as shown in Figure 4.17a and 4.17b.

The center of a large arc could, depending on the scale of the drawing, be "off the paper" when it is printed. The center is specified as shown in Figure 4.17c. The portion of the dimension line touching the arc is a radial line that would pass through the actual center if extended.

It is not good design practice to have sharp corners. A crack will often start at a sharp corner. Sharp corners are eliminated by forming a small radius at each corner, as in Figure 4.17b. This feature is called a **fillet** at an inside corner and a **round** at an outside corner. (It can also be called a fillet at an outside corner.) Dimensioning filleted corners can be done with a note such as ALL FILLETS AND ROUNDS 2 mm.

When an outline consists of several arcs, not all center locations may be specified. Locations of the centers are determined by the tangent points between arcs. Some center locations must be specified so others can be determined. Geometric construction is required to locate intermediate center locations and tangent points, or the SNAP commands on your CAD program can be used. Figure 4.18 shows an example of an outline comprising several arcs. Arc location is controlled by other features; in this case, tangent locations.

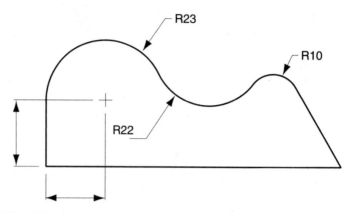

Figure 4.18   Dimensioning an outline composed of several arcs

Now let's move on to look at specification and dimensioning of thread-ed fasteners.

Threaded Fasteners

Threaded fasteners, such as nuts, bolts, and screws, are used to hold parts together. A threaded connection has two components: one with an external thread and one with an internal thread. The internal thread may be in a threaded hole or a nut. The design of threaded connections is beyond the scope of this book: we will deal only with specification and dimensioning of threaded fasteners.

Thread Terms

Figure 4.19 shows thread terms on the cross-section of an external and internal thread.

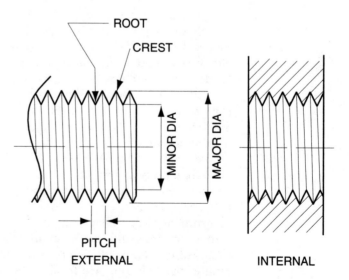

Figure 4.19   Thread terms

The **major diameter** is the outside diameter of an external thread or the maximum diameter of an internal thread.

The **minor diameter** is the diameter at the root of the thread of an external thread and the minimum diameter of an internal thread.

The **crest** is the top of a thread and the **root** is the bottom.

Pitch is the distance from a point on the crest of one thread to the corresponding point on the crest of the adjacent thread.

Other thread terms will be described as required.

Thread Specifications

Some common thread profiles are shown in Figure 4.20.

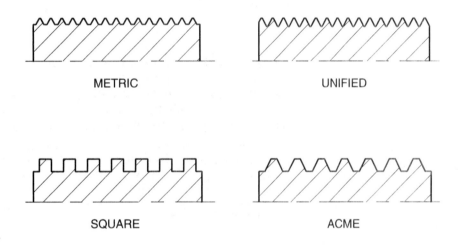

Figure 4.20   Common thread profiles

Metric threads are the international standard thread. **Unified threads** are the standard thread in the USA, Canada, and England. Square, Acme, and buttress threads are used for power transmission.

Since thread dimensions are set by international and national standards, there is no need to give the dimensions. You only need to give the thread series, diameter, length, and the **class of fit** (a tight or loose fit) to specify a thread. There is also no need to draw the thread profile, as shown in Figure 4.20. Threads are represented by a symbol which is easy to draw (see Figure 4.21).

Metric Threads

General purpose metric threads are specified by the major diameter and the pitch. The designation M10 × 1.5, for example, means a 10 mm diameter metric thread with a 1.5 mm pitch.

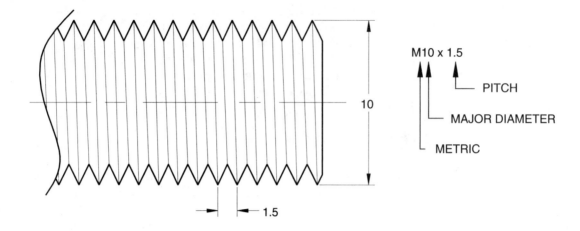

Figure 4.21 Metric thread specification

Pitch and diameter go together so that once the diameter is selected, there are only a few possible pitches. For example, a 10 mm thread can be "coarse" (1.5 mm pitch) or "fine" (1.25 mm pitch). There are other possible pitches for a 10 mm thread, but 1.5 mm and 1.25 mm are the common ones.

A complete specification would include the thread tolerance class. For example, M10 × 1.5 – 6g, where 6 is the tolerance grade and g is the tolerance position symbol (see Figure 4.22).

Further information on tolerance grades and position symbols can be found in ISO Standard 965/1.

Thread length, the distance from the end of the part to the last full thread, can be specified as a note or dimensioned on the drawing.

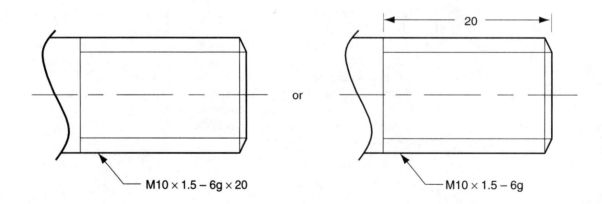

Figure 4.22 Thread symbols used with metric dimensions

The thread is represented by a symbol. The outside diameter is equal to the major diameter and the inside line represents the minor diameter. The line representing the minor diameter is located "by eye" and there is no need to determine the minor diameter (which you must look up).

Unified Threads

A unified thread is specified by the major diameter, number of threads per inch, thread series, whether the thread is coarse or fine, and the class of fit (a tight or loose fit). The notation 1 - 8UNC - 2A specifies a 1-inch diameter, 8 threads per inch (tpi), Unified National Coarse thread, with a class 2A fit.

There are several possible combinations of diameter and threads per inch depending on the thread form; for example, 1 - 12UNF - 2A specifies a 1-inch diameter, 12 threads per inch, Unified National Fine thread, with a class 2A fit.

There are six classes of fit in the unified system. Classes 1A (loose), 2A (standard), and 3A (tight) apply to external threads, and classes 1B, 2B, and 3B apply to internal threads. Classes 2A (external threads) and 2B (internal threads) apply to commercial bolts, nuts, screws, and other fasteners.

In addition to the size and type of thread, length must also be specified, either by a dimension on the drawing or by a note. Since fasteners are standard parts, they are often specified with a note. Figure 4.23 shows how length can be specified.

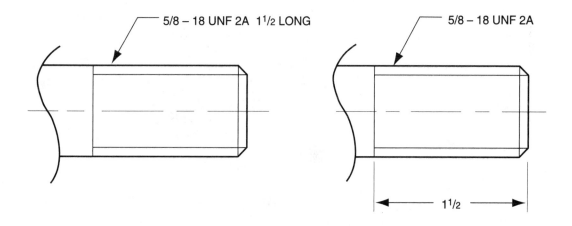

5/8 – 18 UNF 2A 1¹/₂ LONG

5/8 – 18 UNF 2A

1¹/₂

Figure 4.23   Specifying thread length

Two thread symbols are used with the unified system. The thread symbol used in Figure 4.24, called a **simplified symbol**, is similar to that used with metric dimensions. The dotted line represents the minor diameter, which is located by eye. Figure 4.24 shows both thread symbols used with the unified system.

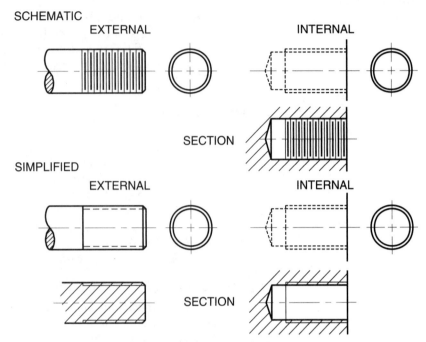

Figure 4.24   Thread symbols used with the unified system

The distance between the lines in the schematic symbol represents the pitch, which, since the number of threads per inch is known, can be found.

Nuts and Bolts

Nuts and bolts are illustrated in Figure 4.25. The heads of nuts and bolts are generally hexagonal, but square heads and square nuts are available.

Figure 4.25   Nuts and bolts

Carriage bolts have round heads. Figure 4.26 illustrates square and hexagonal bolt and nut heads. Dimensions of nuts and bolts are generally not given (except bolt length) since they are standard parts, but dimensions must be known to check clearances and to draw them.

SQUARE HEAD

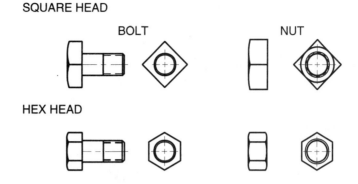

HEX HEAD

Figure 4.26 Types of bolt heads and nuts

Nut and bolt dimensions can be found in Appendix D.

Cap Screws and Machine Screws

Cap screws and machine screws are similar to bolts but have different types of heads. Cap screws and machine screws are generally screwed into threaded holes. Figure 4.27 shows the different head designs available.

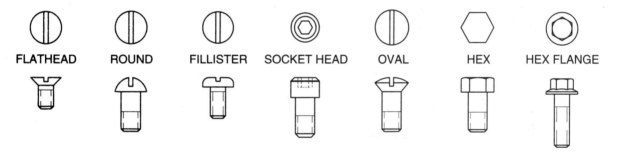

Figure 4.27 Heads for cap and machine screws

Machine screws are smaller in diameter than cap screws and come in sizes from #0 (0.060 inch diameter) to 3/4 inch diameter. Sizes less than 1/4 inch are designated with a number. Cap screws range in size from #0 to 1 1/2 inches diameter.

Standard dimensions can be found in the tables in Appendix D.

Countersunk, Counterbored, and Counterdrilled Holes

These names refer to the shape of the top of a threaded hole, which is often required to join two parts with a threaded fastener without interrupting the surface. The top of the threaded hole must be modified so that the fastener is either flush with the surface or below it. All three holes are shown in Figure 4.28, along with the threaded fastener that would go in the hole.

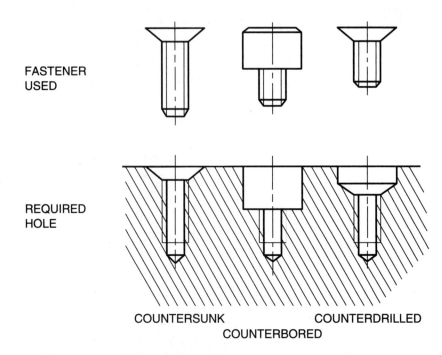

Figure 4.28 Countersunk, counterbored, and counterdrilled holes

Dimensions are given with a combination of notes and symbols, as shown in Figure 4.29.

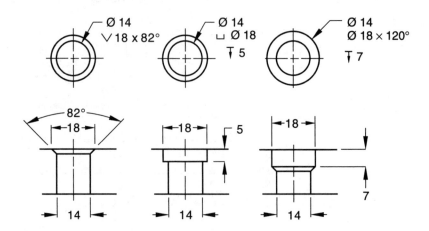

Figure 4.29 Dimensioning countersunk, counterbored, and counterdrilled holes

Spotfaced Holes

In many applications, a bolt head could be above the surface but the bolt head must sit on a flat surface. If the surface is uneven or rough, the bolt will not seat evenly and high loads can result.

The area around the top of the hole is machined to make it flat and smooth. This is called **spotfacing** and the result is a **spotfaced hole**. Figure 4.30 shows a bolt in contact with a sloping surface. Spotfacing the hole allows the bolt to rest on a flat surface. It is similar to a counterbored hole except only a small amount of material is removed and no depth is specified.

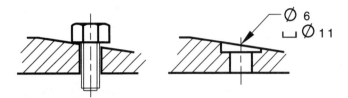

Figure 4.30   Spotfaced hole

Slotted Holes

If the position of a piece of equipment must be adjusted and it is secured with fasteners, the fasteners must be removed for repositioning. If the fasteners pass through slots, any repositioning requires only that they be loosened. Slotted holes usually have rounded ends and straight sides. The length and width of the slot must be specified. Three ways of doing this are illustrated in Figure 4.31.

The radius of the arc on each end is indicated (2X R), but not dimensioned. Specifying width and length as either center distance or overall length gives sufficient information to make the slot.

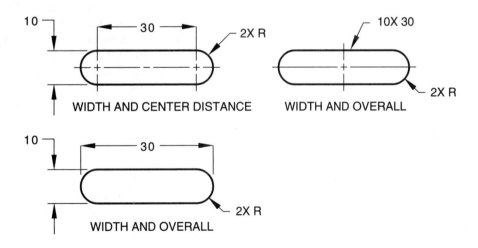

Figure 4.31   Dimensioning a slotted hole

Now let's move on to look at location dimensions.

Location Dimensions

Location and size must be specified to completely dimension a feature. The location of a corner is specified by two linear dimensions. The location of features that are not on the outline are dimensioned in a similar way except that extension lines are different from those used on the outline. Multiple copies are usually dimensioned only once.

Let's look at dimensioning arcs and holes.

Arcs

Arcs are dimensioned on the view in which they are seen as arcs. When the location center is specified, it is identified by a small cross. The position of this cross is specified by extension lines and dimensions. Figure 4.32 shows how the center is defined and specified.

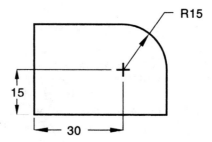

Figure 4.32   Locating the center of an arc

Holes

The location of a hole is generally given on the view in which it appears as a circle. Depth cannot be seen in this view, so it must be specified in a different way or dimensioned on another view (but not as a hidden line). Figure 4.33 shows how to dimension a hole. The location of the center is specified by extension lines that are extensions of centerlines. There is no gap between the centerline and extension lines. Extension lines at the other end of the dimension line have a gap between the line and the outline.

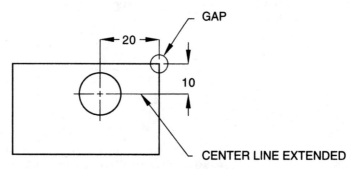

Figure 4.33   Dimensioning the location of a hole

Now let's move on to look at how to dimension repeated features.

Repeated Features

It is common to use one hole size or other feature more than once in a part. If four holes are needed in one part, they would usually be made all the same size, unless there was a good reason to use different sizes. The same tool can then be used to make all of them and set-up time and costs are reduced.

When a feature is repeated, there is no need to put the same dimension on each one. Size dimensions and sometimes location dimensions are specified only once and the number of repetitions is given. Figure 4.34 shows an example with repeated features, holes and radii, with a common center.

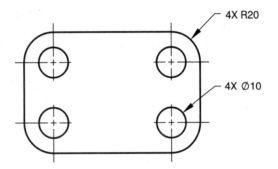

Figure 4.34   Dimensioning repeated features

The symbol 4X Ø10 means there are four holes with a diameter of 10 mm, and 4X R20 means there are four corners with a radius of 20 mm. It should be obvious by looking at the drawing where these features are.

If some feature is repeated at a regular spacing, there is no need to dimension every length or draw the feature many times, although this is easy to do with a computer program. The same circle drawn a hundred times takes more time to draw and increases plotting time. Figure 4.35 shows how an array (rows and columns) is dimensioned.

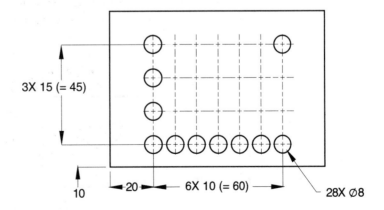

Figure 4.35   Dimensioning arrays

The drawing indicates that there are four rows and seven columns of 8 mm diameter holes The holes are spaced 15 mm apart vertically and 10 mm apart horizontally.

The same feature can be repeated in a circular direction. In this case, angular dimensions may be more convenient than linear ones, although linear dimensions could be used. Figure 4.36 shows an array of holes arranged in a circle.

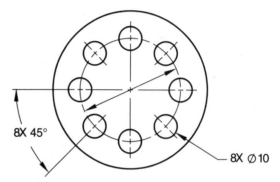

Figure 4.36 Angular dimensioning of a repeated feature

The same symbols are used to specify the number of repetitions and the spacing.

8X Ø10 means eight holes with a diameter of 10 mm and
8X 45° means eight spaces of 45° (the holes are spaced at 45°).

It is common to locate the center of each hole at the same radial distance. This distance is indicated by a centerline and radial lines are drawn at some, or all, of the hole positions. The circle on which holes are located is sometimes called the **bolt circle diameter** (BCD), assuming bolts are used in the holes. At least one hole must be drawn to indicate the starting point. Circular arrays can be drawn easily with a computer program.

Linear dimensions can be used instead of radial dimensions to locate holes or other circular features, as shown in Figure 4.37.

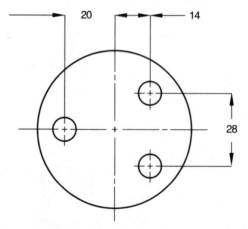

Figure 4.37 Dimensioning radial features with linear dimensions

Now that you know a little more about dimensioning features, threaded fasteners, location dimensions, and repeated features, try the following problems.

Problems

Dimension the following objects in accordance with the guidelines given in this chapter. Dimensions are found by scaling the drawings and rounding dimensions to the nearest millimetre, or as specified by the instructor. The scale must be specified on the drawing.

The finished drawing (or sketch) can be passed to another member of your work group for checking. The "checker" marks all errors and omissions and returns it to the originator for correction.

26.

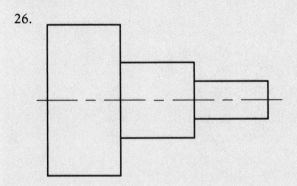

27.

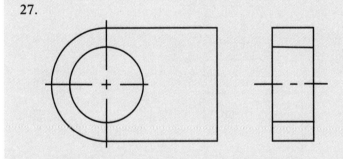

28.

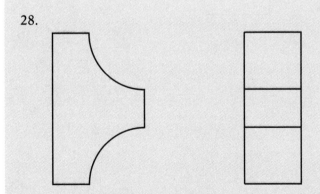

29.

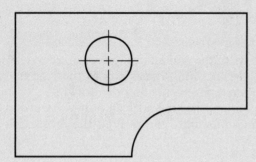

30.

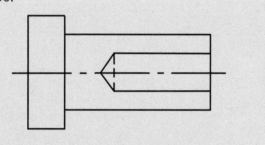

31.

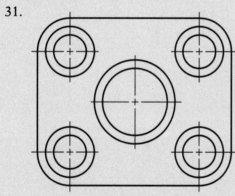

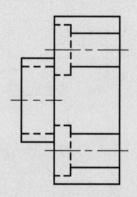

32.

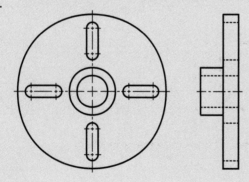

33.

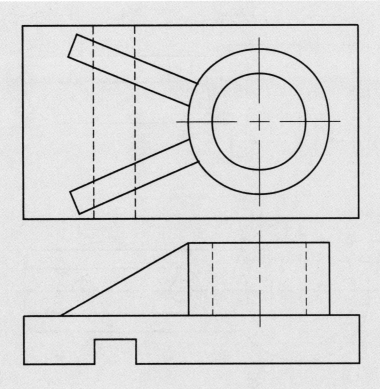

34.

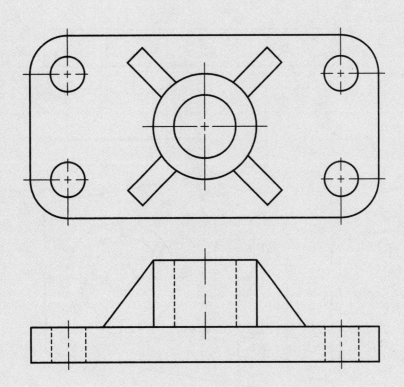

35.

36.

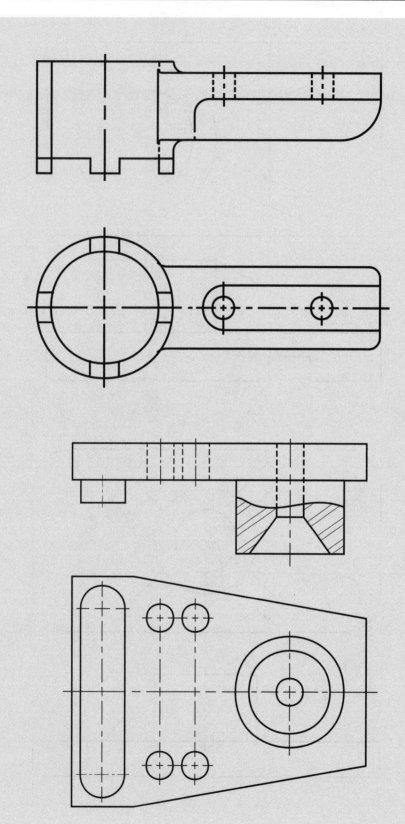

37.

38.

39.

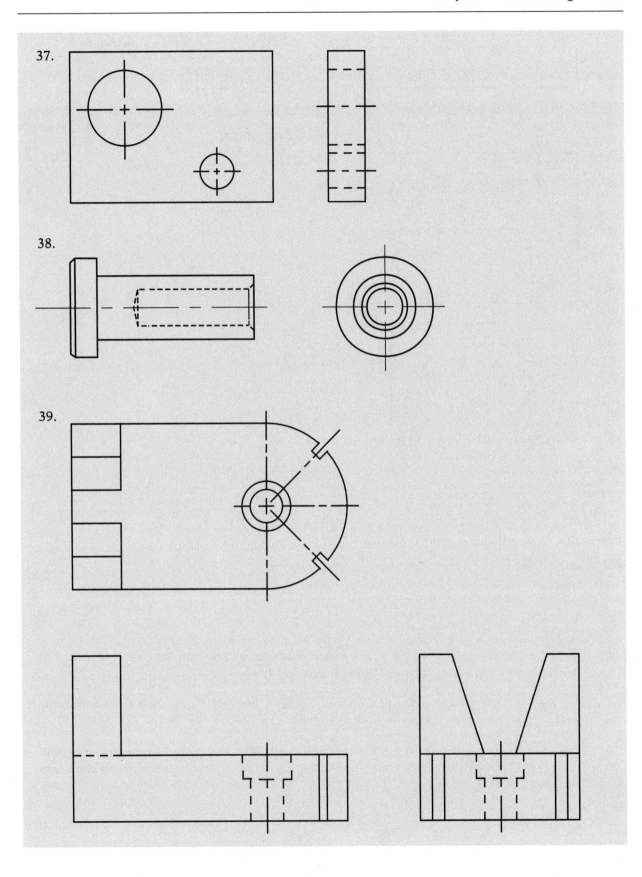

40.

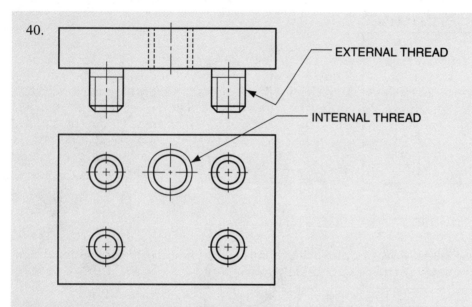

EXTERNAL THREAD

INTERNAL THREAD

41.

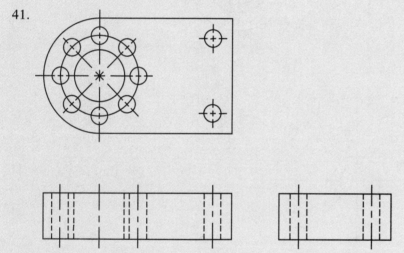

42. Create a dimensioned drawing of a 3/4 inch diameter by 250 mm long shaft threaded at both ends with a coarse thread. Thread length is 50 mm.

43. Create a dimensioned drawing of a 20 mm diameter by 8 inch long shaft threaded at one end with a fine thread. Thread length is 2 inches.

44. Detail the end connections for a W 610 x 140 beam. The end plate beam connectors will have 4-3/4 bolts arranged vertically. You will have to look up the dimensions of the beam and connector.

Scales

Drawings are drawn "full size" with a computer but they cannot be plotted full size if the object is large (an airplane, for example). A reduction scale, specified by the user, is applied when the drawing is plotted. One unit drawn on paper represents a larger number of units on the real object. For example, 1 mm on the paper represents 100 mm on the real object. Dimensions shown on the drawing are actual dimensions.

Scales used on engineering drawings are not arbitrary: only certain ones are used. The scale must always be stated on the drawing.

Metric Scales

Metric scales used on engineering drawings are specified by national standards. Metric scales are expressed as:

1:1	1 mm on paper = 1 mm (this is full size)
1:2	1 mm on paper = 2 mm
1:5	1 mm on paper = 5 mm

Larger ratios follow the same sequence (1, 2, 5):

1:10	1:100	1:1000	1:10000
1:20	1:200	1:2000	1:20000
1:50	1:500	1:5000	1:50000

A metric scale will typically have 1:10, 1:20, 1:50, 1:100, 1:200, 1:500, and 1:1000 scales. Other scales can be used by simply adding or subtracting a zero.

When an engineering drawing is not done with a computer, it is drawn at the desired scale. A special ruler, called a **scale**, is used to lay out lines. Length can be read directly from the scale so no calculations are required. Figure 4.38 shows a typical metric scale.

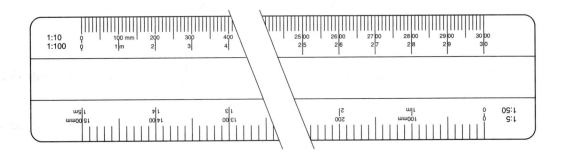

Figure 4.38 Typical metric scale

A length is marked off by reading the actual length on the scale and marking the end point. Figure 4.39 shows how lengths of 16 400 mm at a scale of 1:200 and 43.5 m at 1:500 are laid out. The desired length is read on the appropriate scale and marked on the line.

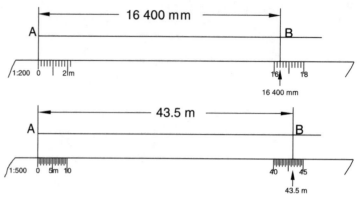

Figure 4.39   Laying out a length with a metric scale

In the past, other scales were used on engineering drawings and many drawings with these scales still exist, so you should be familiar with them. They are named after the branch of engineering in which they are commonly used. The civil engineer's, mechanical engineer's, and the architect's scale are fully described in Appendix E.

To help you understand the practical application of your knowledge, take a few minutes to look at the following examples.

Example 4.1

The parts shown in Figure 4.40 are made by cutting a slot in a 50 x 100 workpiece with a 3 mm thick cutter. Slots are made by passing the blank under a cutter. The table and stop are adjusted to locate the blank in the correct position relative to the cutter. The table can be raised and lowered and the stop moves left or right.

A setup gauge is required to set the position of the stop and the height of the table. Specify a simple gauge that can be used to set the depth of cut and slot location for either part. One gauge will be used for both parts.

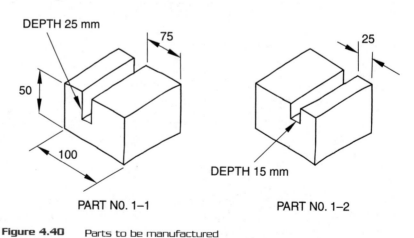

Figure 4.40   Parts to be manufactured

What is known?

- Each workpiece is 50 mm high × 100 mm wide.
- Slots are required at 25 mm and 75 mm from one edge. These distances are to the edge of the slot.
- Slots are 25 mm and 15 mm deep respectively.

The setup is shown in Figure 4.41.

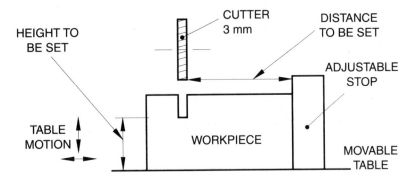

Figure 4.41   Setup of workpiece and cutter

Figure 4.42 shows how a gauge could be used to set the table and stop. The setup gauge is shown as a dashed line for each setup.

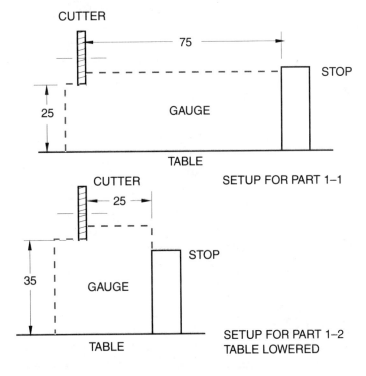

Figure 4.42   Sketch showing how gauge is used

A dimensioned drawing of a gauge to handle both parts is shown in Figure 4.43.

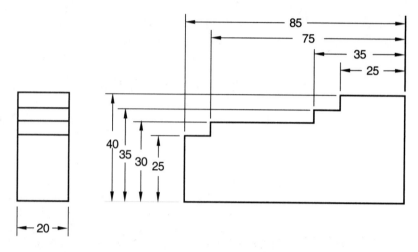

Figure 4.43 Finished gauge

The thickness (20 mm) is not critical; it was chosen to be large enough for the gauge to stand upright. The height (40 mm) and width (85 mm) are likewise not critical; convenient values were chosen.

The final drawing should be checked to see if there is information missing. Ask yourself if the gauge could be made with the information given.

Example 4.2

A manufacturing operation requires that mixtures be mixed at three different speeds—100, 50, and 25 rpm. A motor running at 50 rpm is available. The motor has a 50 mm diameter pulley for a 20 mm belt drive. The mixer has a 12 mm diameter shaft.

The motor cannot be connected directly to the mixer because it turns too fast, so a speed-reducing pulley is required on the mixer shaft. Specify, with a detailed drawing, a suitable speed-reducing pulley for the mixer shaft.

What is known?

- The drive motor turns at 50 rpm and has a 50 mm diameter pulley.
- The mixer must operate at 100, 50, and 25 rpm.
- The mixer shaft is 12 mm diameter.
- The drive belt is 20 mm wide. Figure 4.44 shows this information.

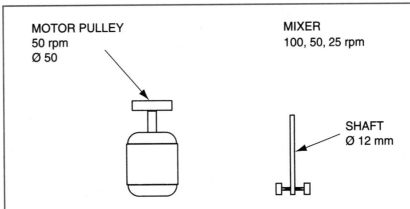

Figure 4.44   Sketch showing given information

If the mixer pulley (the driven pulley) is the same size as the motor pulley (driving pulley), the speeds will be the same; if it is larger, mixer speed is reduced. Since three speeds are required, three different-sized pulleys must be used.

Figure 4.45 shows a sketch of the proposed solution. No supporting structure is shown. Speed is changed by moving the drive belt to a different-sized pulley. Pulley diameters on the mixer can be determined knowing the required mixing speeds.

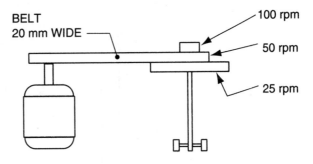

Figure 4.45   Proposed pulley arrangement

The ratio of mixer speed to motor speed depends on the ratio of motor pulley diameter to mixer pulley diameter and is given by:

$$\text{MIXER SPEED} = \text{MOTOR SPEED} \times \frac{\text{MOTOR PULLEY DIA}}{\text{MIXER PULLEY DIA}}$$

Required diameters are:

Driver speed	Mixer speed	Required diameter
50 rpm	100 rpm	25 mm
	50 rpm	50 mm
	25 rpm	100 mm

Figure 4.46 shows a sketch of the system with the new size pulleys on the mixer. (The details of how this pulley will be attached to the shaft will not be considered now.)

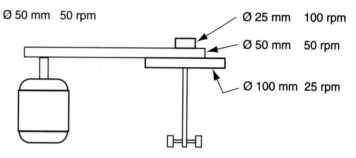

Figure 4.46   Drive arrangement and pulley sizes

Figure 4.47 is a drawing of the mixer pulley showing how all diameters and holes are dimensioned. This drawing shows only the information on diameters. Other information and dimensions must be specified before the pulley can be made. For example, how is the pulley attached to the mixer shaft?

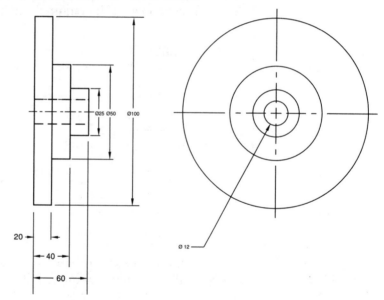

Figure 4.47   Pulley diameters specified

Pulley diameters are shown on the view in which both diameter and length are seen. The hole for the mixer shaft is specified on the side view where it shows as a circle. This hole has been specified as 12 mm (the same as shaft diameter) so it will be a tight fit on the shaft.

Now try the following design problems.

Problems

The following problems are open ended in that there is more than one possible solution. Requirements are stated in general terms and are subject to interpretation. You will have to decide some of the criteria. You will be required to do some research on your own to determine some of the variables (standard sizes, material weights) in order to fill in missing information. Some, but not all, design questions are posed for each problem.

Prepare a drawing, or drawings, either by hand or computer, so that the project could be made.

45. Design a portable holder for a flag staff. The holder will rest on the ground and hold a flag staff 2 m long and 50 mm diameter at the base. The flag staff must be easily removed from the holder and the holder must be moved by one person. The flag staff must be held in a vertical position and must not fall over under normal conditions.

    Since the base must be moved by one person, how much can an average person carry? The angle at which the flag staff can be tipped before falling depends on the location of the center of gravity and the size of the base. How could it be made stable?

46. Design a bookcase to hold textbooks, computer manuals, and other items you use. The bookcase must not take up too much floor space and must be easily moved at the end of the term. Shelves should not sag when loaded. The bookcase must be stable and not tip over.

    How long can a shelf be before it sags under a typical load of books? What are the sizes of the books that will be put in this bookcase? If you live in a region of potential earthquakes, should something be included to prevent books from falling off the shelves or the bookcase from falling in the event of an earthquake? How much will it cost? How much will it weigh?

47. Design a device to pour liquid from a 45 gallon steel drum. The drum must be lifted and tilted so that all liquid in the drum can be poured into a 1500 mm high tank. The liquid in the drum has a density 10 percent greater than water. The lifting/tilting device must be easily attached and removed from the 45 gallon drum by one person. A crane is available for lifting the drum. The unit must be easily and safely operated by one person. What are the dimensions of a 45 gallon drum and how much will it weigh when full?

48. Design a device to pick up a 500 mm long by 150 mm diameter cylinder from the bottom of a 5 m deep concrete tank filled with clear water. The cylinder contains uranium and is radioactive. It must be kept at least 3 m below the surface. The lifter must be operated by one person. How much does the cylinder weigh? What will the lifting device weigh?

49. Design a support system for an elderly person to use when entering or leaving a bathtub.

50. A patio restaurant has a sliding roof that can be rolled into place when it rains. The roof section rolls on wheels but because there is nothing to guide the wheels, it is difficult to roll the roof in a straight line and it is difficult to close. It is proposed to replace one or possibly two wheels with new wheels that roll on some guiding system so that the roof will always roll along the same path. Design a new wheel and guide system to do this. The restaurant owner is very concerned with cost. The roof must be operated by hand by unskilled persons. The existing axles are 25 mm diameter and will accommodate a wheel 50 mm thick.

51. Design a support to hold a computer monitor from a wall so that it can be at a convenient height for the user and does not take up space on the desk. How much does the average monitor weigh? What features could be added to improve the bracket?

52. Design a support to hold a printer under a table. The bracket must be supported from the table top and not from the floor. Printer controls and paper must be easily accessible by a person sitting in a chair.

53. Design a device to hold a bicycle when doing maintenance and repairs at home. The device would be held in a vise or rest on the floor and be easy to operate by one person. It should be capable of firmly holding a bicycle in different positions.

54. Design a simple easel that could be used by children 3 to 5 years old when they paint or color at home. Incorporate some method of holding a paper supply that can be used by the child. The easel should fold easily for storage.

55. Design a play table for children to play with small toys. The age range of the children is 4 to 7 years and the children may sit or stand at the table. Bear in mind that children may climb on the table, so it must be strong enough to allow for this.

56. Design a device that will lift a 50 kg crate from the floor to a height of 1 m where it is slid onto a roller conveyer and taken away. The device must be easy to operate and the crate must be easily slid onto the roller conveyer. The crate is 300 mm wide, 450 mm long, and 230 mm high.

57. It is often useful to have extra light in a localized area; for example, when painting. Design a fixture that will hold a standard two-tube fluorescent light fixture. The fixture should be able to hold the light so it illuminates a horizontal surface, such as a table or a vertical surface. The fixture should be easily movable and take up as little space as possible.

58. Design a support that will hold a personal computer under a table. The holder should provide security against theft and protection in the event of an earthquake. All controls and cable connections must be easily accessible.

59. Design a method of measuring the level of water in a rectangular tank. The tank is wood and no holes of any sort can be put in the walls. A wooden cover is used and, since it is always above the water level, small holes can be made in the cover. Anyone wanting to see what the water level is must be able to do so 6 m from the tank. The tank is in a remote area without electricity, and no battery-powered devices can be used. Assembly and maintenance must be done with simple tools by unskilled workers.

Chapter 5

Engineering Drawings

Engineering drawings, sometimes referred to as **technical drawings**, are used to communicate design, project, and production information, under the direction of an engineer.

Engineering covers many different disciplines; for example, mechanical, mining, metallurgical, structural, civil, environmental, geotechnical, electrical, electronic, industrial, petroleum, and nuclear to name a few. Engineering drawings that are used to supply information and instructions for the manufacture of products or the construction of a structure, are called **working drawings** or **production drawings**.

Working Drawings

Working drawings are a complete set of documents specifying all of the information required to make or build the design. Working drawings may be on one or more sheets, depending on the size and complexity of the design, and may include written specifications or instructions.

A working drawing is a legal document and should be drawn in accordance with the relevant national or international standards governing the particular discipline. Many companies adopt their own standards that are generally based on the national standards.

Examples of national standards are:

Canada	CSA B78
USA	ANSI Y14
Britain	BS 308

National standards are internationally referenced by the International Standards Organization (ISO) in Geneva, Switzerland. The ISO standard for technical drawings comprises some 71 international standards grouped under:

ISO/TC 10, *Technical Drawings, product definition and related documentation*.

In most cases engineers do not create working drawings, although in a small enterprise this could happen. Usually they leave the drafting to technicians or drafters. However, it is essential that engineers are familiar with the standards as they (the engineers) are ultimately responsible for the content and correctness of the documents, and generally stamp them accordingly. Engineers, therefore, must be able to interpret what is drawn and be able to give instructions for design changes (**engineering change orders**), if necessary.

Working, or engineering, drawings must be clear as to the designer's intent and must contain all of the information necessary to manufacture or build the design. Any misunderstanding or ambiguity when reading the drawings can cause expensive changes during production or building. This, in turn, wastes time and money. Because it is a legal document, the information on the working drawing must be accurate. If there is any failure to the design or structure, people will first look at the information on the working drawing, from which the design was implemented.

Working drawings can take several different forms, depending on the discipline; for example, in mechanical, civil, chemical, and electrical/electronic engineering, engineering drawings can be generalized for content-discussion purposes. First we will discuss some items common to all working drawings.

Drawing Sheets

Whether the finished drawing is done by hand or plotted from a CAD (computer-aided design) system, working drawings should be produced on standard-sized sheets. The following table gives the standard sheet sizes for technical drawings:

Metric (millimetres)	Inch (standard)	Inch (civil/architectural)
A4 – 210 × 297	A – 8.5 × 11	9 × 12
A3 – 297 × 412	B – 11 × 17	12 × 18
A2 – 420 × 524	C – 17 × 22	18 × 24
A1 – 594 × 841	D – 22 × 34	24 × 36
A0 – 841 × 1189	E – 34 × 44	36 × 48

It is worth noting that the drawing media used can make a difference to the quality of the output. While manufacturers or builders rarely see plotted or drawn "originals," they do see copies of the original, reproduced by either a photocopy or diazo whiteprint process, or from a microfilm negative. In most cases, some of the original quality; i.e., line sharpness and text legibility, is lost. The engineer "signing off" the drawings must ensure that the drawings are properly prepared and leave no room for misinterpretation.

The most popular forms of drawing media are opaque paper (bond), translucent paper (vellum), and polyester film (mylar). Film is the most expensive media, and the most stable for clarity. It is also water- and heat-resistant. It is recommended where original drawings are to be kept for a long time and handled or modified frequently. Opaque paper is normally used for master documents such as maps and charts. Engineering production and construction drawings are plotted or drawn on translucent paper or film. Drawings on opaque paper are reproduced using an electrostatic process such as photocopying. Drawings created on translucent paper and film can be reproduced using photocopying or the less expensive diazo process.

Drawing Sheet Format

Every production or construction drawing sheet will include an area on the right-hand side containing a title block and any general notes pertaining to the drawing. In addition, drawing sheets should include a grid reference zoning system to locate changes, and centering marks to facilitate positioning of the sheet when microfilming, as shown in Figure 5.1.

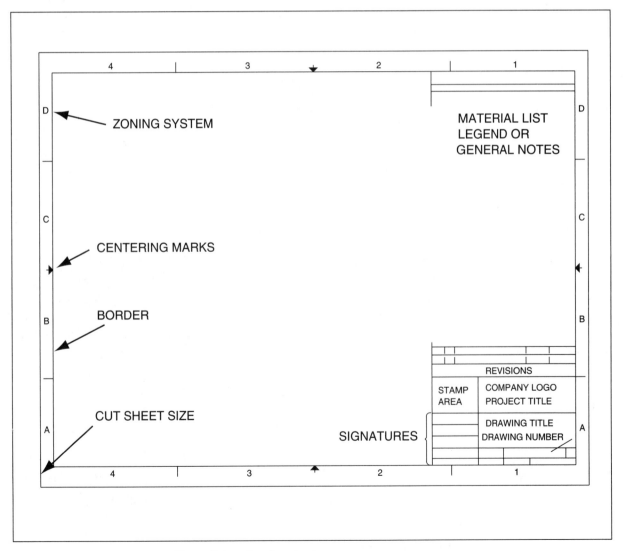

Figure 5.1 Drawing sheet

6			
5			
4			
3			
2			
1	SEPT / 98	RDW	REVISED PUMP INSTALLATION
REV.	DATE	BY	DESCRIPTION

PEA Professional
Engineering
Associates

CLIENT
PULP CONSOLIDATED
LIMITED

PROJECT TITLE
THICKENER OVERFLOW TO
POWER BOILER SCRUBBER

DRAWING TITLE
PLAN AND DETAILS
MECHANICAL

DATE	JAN./98	STAMP
CHECKED	S. O'HALLORAN	
DRAWN	B. Andrews	
DESIGNED	C. Black	
SURVEYED	-	
SCALE	AS SHOWN	

		JOB NO.
APPROVED _____ 19 _____		9802
		SHEET
APPROVED _____ 19 _____		1 OF 2
DRAWING NO. 9802-M01		REV. 1

Figure 5.2
Title block

Title Blocks

Title blocks vary in layout and contain general information relating to the drawing, such as the company's name, project title, drawing title, date drawn, scale used, and drawing number. Figure 5.2 is an example of a consultant engineering title block. Notice that the client's name is included in addition to the consultant's name. Complex parts may require several sheets to fully describe the part or structure; therefore, sheet numbering is added to indicate the number of sheets in the drawing set; e.g., sheet 1 of 2.

The project title is the general title of the design; e.g., THICKENER OVERFLOW TO POWER BOILER SCRUBBER. The drawing title is for a particular drawing description; e.g., PLAN AND DETAILS— MECHANICAL.

Most firms have their own title block already printed on cut sheets or stored in a CAD database for plotting. For presentation purposes, it is important that all drawing sheets in a set are consistent throughout; i.e., the same size, the same title block text. Auxiliary drawing numbers may be added in the upper left corner, usually to read from the top-down, to aid hard copy (paper) filing systems.

Provision is usually made in a title block for approval signatures and professional stamps. Manufactured parts' information such as general tolerances, material used, projection symbols, and drawing standards may also be included within the title block.

Companies using CAD store pre-drawn title sheets and use them whenever they start a new drawing. CAD programs usually come with pre-drawn title sheets, as templates, which you are encouraged to make use of for projects in this book.

Revision Column

An important area of the title block is the **revision column** or **change table**. This area contains a brief description of any changes made to the drawing, along with the dates of the changes and approval signatures. Any changes to the drawing after it has been issued for manufacture or construction must be recorded and approved. The area containing the change can be listed using the grid reference or often a "cloud" is drawn around the area with the revision number indicated. Figure 5.3 shows a change to a dimension.

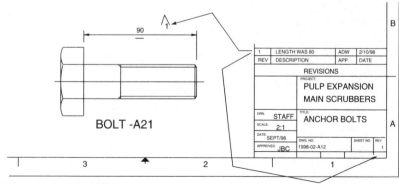

Figure 5.3 Revision column

NOTES

1. CONTRACTOR TO VERIFY ALL DIMENSIONS IN FIELD BEFORE FABRICATION OF PIPE.
2. SEE PUMP SPECIFICATIONS FOR DETAILED DESCRIPTION OF PUMP.
3. CARBON STEEL PIPE SUPPORTS AND STAINLESS STEEL PIPE TO BE SEPARATED BY A RUBBER WRAP. RUBBER TO HAVE 1" OVERHANG ON EACH SIDE AND BE NEATLY WRAPPED AROUND SUPPORT AND SEALED IN PLACE.
4. MAXIMUM DISTANCE BETWEEN PIPE SUPPORTS TO BE 10 FEET.

LEGEND

- - -✳- - - TIE-IN POINT (NEW PIPING TO EXISTING)

⊢⤂ KNIFE GATE VALVE, STAINLESS STEEL

(PSS) PIPE SUPPORT TAG (e.g. PIPE SUPPORT DETAIL No. 5)

Figure 5.4 Material list/legend area showing general notes and legend

Material List/Legend Area

The area above the title block should be reserved for listing parts or materials, general notes pertaining to the entire drawing, or a legend describing the symbols used on the drawing (see Figures 5.1 and 5.4).

Drawing Sheet Layout

Drawing sheets should be laid out in accordance with good drafting practice. This includes the arrangement of views, angles of projection, and clarity of presentation. The engineer who signs off the drawing must ensure that it is correct and well presented. Remember, the quality of a drawing reflects on both the engineer and the company.

Drawing Views

As discussed in earlier chapters, the North American standard method of laying out multiple views in a 2-D space is to use third angle projection. This includes orthographic, section, partial, enlarged, and auxiliary views. As the intention of the drawing is to communicate, the drawing should be clear, with views spaced apart and dimensions and notes placed in appropriate locations. On an engineering drawing, the number of views should be limited to the minimum necessary to fully describe the design without ambiguity.

Let's look at the various types of views in more detail, starting with orthographic and section views.

Orthographic and Section Views

The principles of orthographic projection are discussed in Chapter 1. The third angle projection system practiced in North America places "flat" or projected views parallel to each other as shown in Figure 5.5.

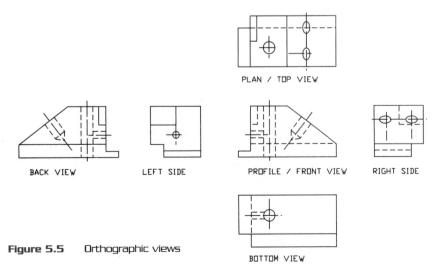

Figure 5.5 Orthographic views

The number of views needed to describe objects or structures for production depends on the complexity of the objects. The two important characteristics of engineering drawings are that they

1. provide sufficient "flat" views to fully describe features (holes, edges, cut-outs, etc.) of the object
2. provide sufficient dimensions to make the object.

Bear in mind the placement of dimensions as discussed in Chapter 4. Edges that are hidden from view should not have dimensions. Place dimensions on the view that best describes the feature.

Section views are used to show interior detail and are discussed in Chapter 3. Using section views in place of regular orthographic views can save drawing time and the number of drawing sheets used. The shaft hanger shown in Figure 5.6 makes use of a full section, in place of the traditional front view, to illustrate the cut-out and recess information, and a revolved section to describe the rib profile.

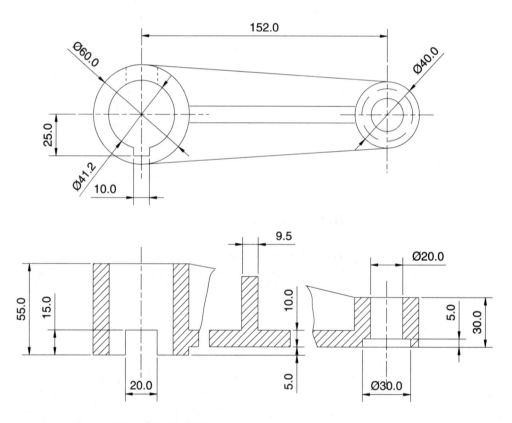

Figure 5.6 Section views

Partial and Enlarged Views

Often it is not necessary to show a complete view: A partial view will suffice. A **partial view** is used to aid clarity and where drawing the complete view would not improve the information given. A symmetrical object can be adequately described by showing only one half or a quarter of the whole view (see Figure 5.7). The line of symmetry is indicated by a center line with two short parallel lines drawn across it. When using CAD, it is important to adhere to these guidelines as it is often easier to give a complicated full view from the 3-D model rather than the required simpler one.

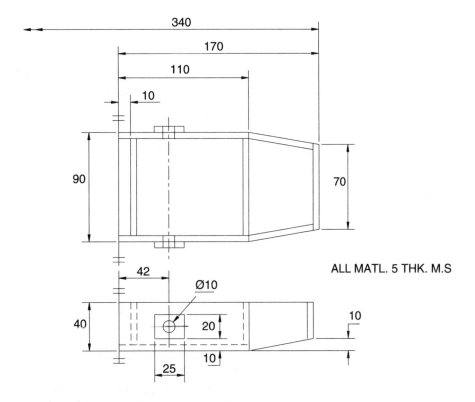

Figure 5.7   Partial view of symmetrical object

Sometimes parts are so complicated, or structures so large, that full views or partial views have to be drawn on separate sheets. The arrows indicating the viewing or cutting plane will include the sheet number where it is drawn.

An **enlarged view** is used to show a partial view or feature detail in an enlarged scale. This is done to prevent overcrowding of dimensions or to simply save time and/or improve clarity. The view is titled and has the scale underneath. If you are using CAD, the entire drawing is created full size in the model or work space, and the views scaled, cropped, and arranged using the paper layout space.

All views should be plotted to a standard scale; i.e., as described in Chapter 4. If the scale is the same for the entire drawing, it will be noted in the title block. If the drawing sheet contains partial views or details drawn at different scales, then the scale will be noted underneath the view title. Figure 5.8 gives an example of this.

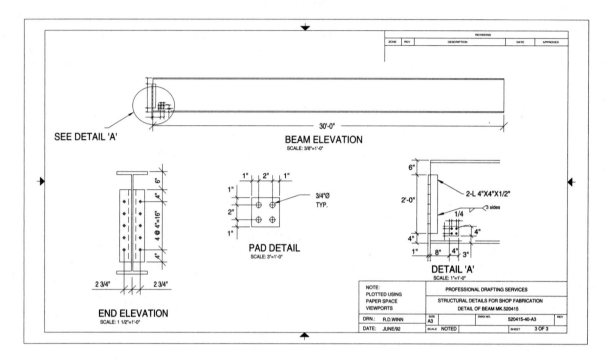

Figure 5.8   Multiple scaled plot sheets

Auxiliary Views

Figure 5.5 shown previously illustrates the six principal orthographic views, as discussed in Chapter 1. Note that the inclined surface appears shortened in both the top and right side views, leaving no place to correctly dimension the holes. Inclined surfaces are only perpendicular to one viewing plane, in this case the front plane.

When it is necessary to show the true size and shape of an inclined surface, you project a view parallel and perpendicular to the inclined surface. This view is called an **auxiliary view** and is normally drawn adjacent to the inclined surface, although it may be drawn anywhere as a removed view.

In Figure 5.9 an auxiliary view has been drawn, enabling the holes in the inclined surface to be dimensioned. Only the true size and shape of the inclined surface are shown, not the entire projection. When only part of a view is shown, it is known as a partial view. Partial views are often used for auxiliary views. A full auxiliary view would not help to describe the features and should be avoided. The top view is also shown as a partial view, with a break line cutting off the unnecessary part of the view. The right side view is also omitted as it serves no purpose.

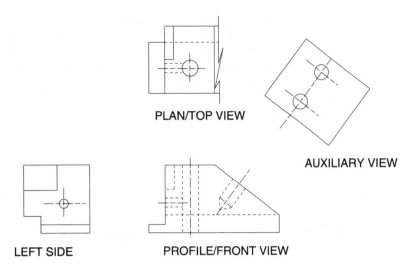

PLAN/TOP VIEW

AUXILIARY VIEW

LEFT SIDE   PROFILE/FRONT VIEW

Figure 5.9   Auxiliary view

In the case of oblique surfaces—i.e., surfaces that are not perpendicular to any of the three standard viewing planes—a second auxiliary view must be projected from the first, or primary, auxiliary view. Figure 5.10a shows a bracket that has an oblique surface. Figure 5.10b shows the fully dimensioned working drawing for the bracket. The secondary auxiliary view is projected parallel to the primary auxiliary view.

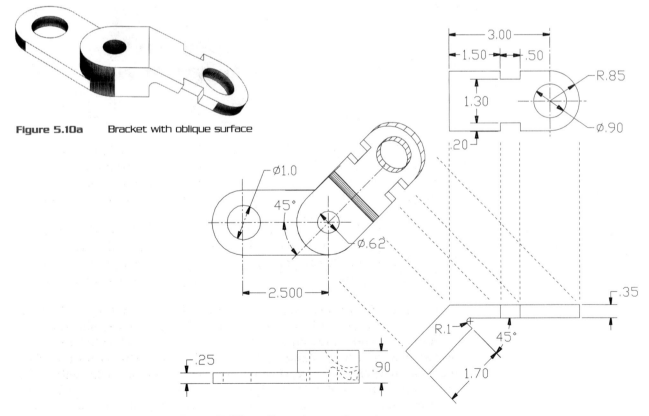

Figure 5.10a   Bracket with oblique surface

Figure 5.10b   Secondary auxiliary view

Tabulated Drawings

Another time-saving tactic used on working drawings is to make a standard drawing and include a table of dimensions for similar parts or arrangements. This is called a **tabulated drawing**. In Figure 5.11 some of the dimensions have been replaced with letters. The values for these are found in the table next to the particular part variety. The drawing here is obviously not to scale for more than one part.

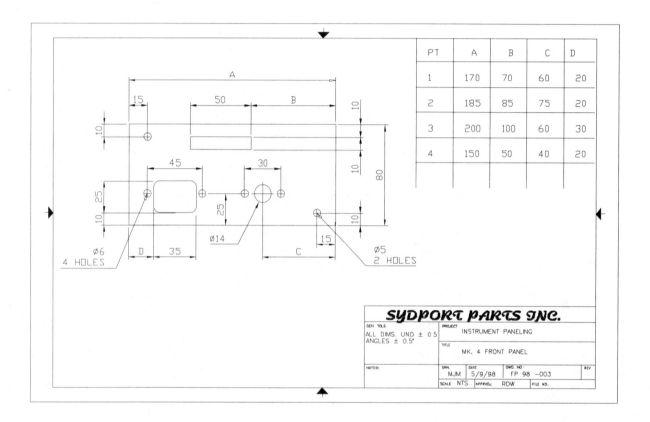

PT	A	B	C	D
1	170	70	60	20
2	185	85	75	20
3	200	100	60	30
4	150	50	40	20

Figure 5.11 Tabulated drawing

Mechanical Engineering Drawings

Mechanical engineering covers a multitude of different types of engineering drawings. Different companies have different meanings for the term **mechanical**. For example, in manufacturing, mechanical engineers design mechanisms and parts, whereas in consulting engineering the mechanical engineer is responsible for heating systems and ventilation, sizing and routing of ductwork, and plumbing systems. The following table lists the general headings along with a description of the working drawing contents.

Mechanical Engineering	Working Drawings Content
Manufacturing	Product parts and assemblies
Building Services	Plumbing, heating, and air conditioning
Material Handling	Conveying, equipment, and supports
Industrial Piping	Routing and supporting pipes
Fluid Power	Hydraulic and pneumatic applications

Manufacturing

The concepts of engineering graphics are taught using examples from product manufacturing. Small parts can require all of the principal views, projection methods, and dimension methodology, without having to contend with the concepts of large-scale drawings.

Manufacturing working drawings are divided into two main types: **detail drawings**, that contain all of the necessary information to manufacture the parts and **assembly drawings** that provide information for the assembly of the parts.

Detail Drawings

A detail drawing is a complete description of a part. This includes shape description, size description, and general specifications. The drawings used so far in this text for examples and problems are detail drawings.

Shape description refers to the views used to describe the part. Orthographic views, auxiliary views, sections, and pictorial views, as described elsewhere in this text, may be used.

Size description refers to dimensioning and tolerancing of the part features. To some extent the manufacturing process and part function will determine the method of placing dimensions and assigning tolerances. The cost of the part can be greatly reduced by decreasing the number of digits to the right of the decimal point!

General specifications include such items as the material used, surface finish, heat treatment, type of finishing process required (e.g., chromium plate), or applicable standards to be used.

Detail drawings may be drawn as one part per sheet (see Figure 5.12) or with several parts drawn on one sheet (see Figure 5.13). One part per sheet is the preferred method for production drawings, as parts may be used on several designs. For complicated parts, several sheets may be required to give a complete description for manufacture. The drawing must be complete and not leave anything open to misinterpretation. Figure 5.14 shows the use of an enlarged detail view to describe features more clearly.

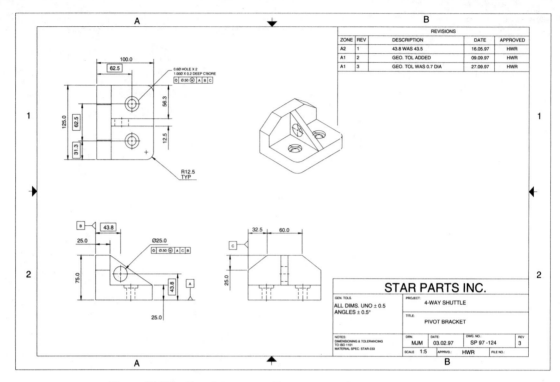

Figure 5.12 Detail drawings—One part per sheet

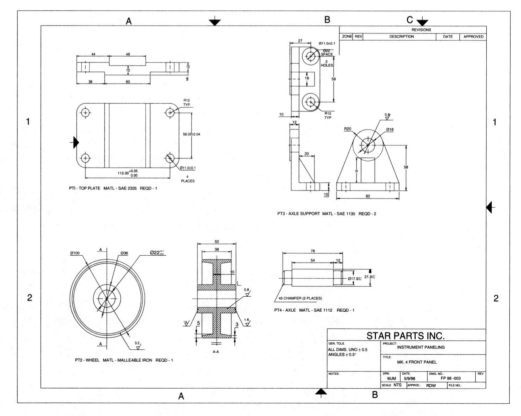

Figure 5.13 Detail drawings—Several parts per sheet

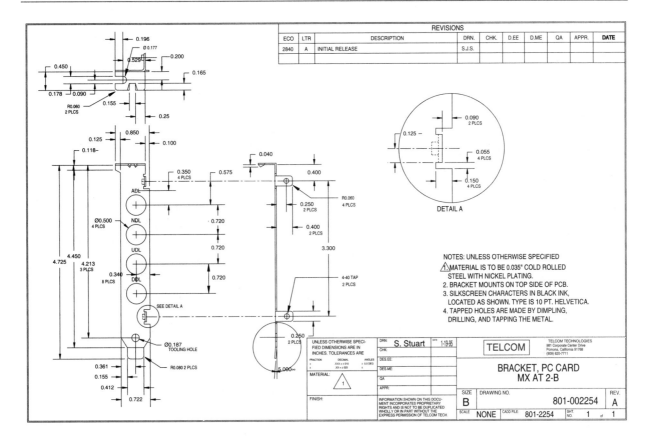

Figure 5.14 Enlarged views [Courtesy: Telcom]

Standard Parts

The number of detail drawings for a product can be reduced (and hence the time taken to get the design to production) by making use of standard parts wherever possible. Standard parts are purchased from a supplier and may include such items as gears, cams, bolts, screws, keys, handles, etc. These items are then listed in a materials list on an assembly drawing. The design engineer may need to see a purchase part drawing or shop drawing for bought-out parts. This overcomes the problem of the supplier making changes unknown to the user.

Fabrication Drawings

Some component parts are fabricated from pieces of steel welded together, called a **weldment**. Automobile frames and bodies, machine tools, and mining equipment all use fabricated components as well as machined parts. These detail drawings will include information on the type and size of weld used, in addition to the items specified above. Often each weldment has all of the separate pieces of steel listed on the right-hand side of the drawing in a material or item list, to aid procurement. Figure 5.15 is an example of this.

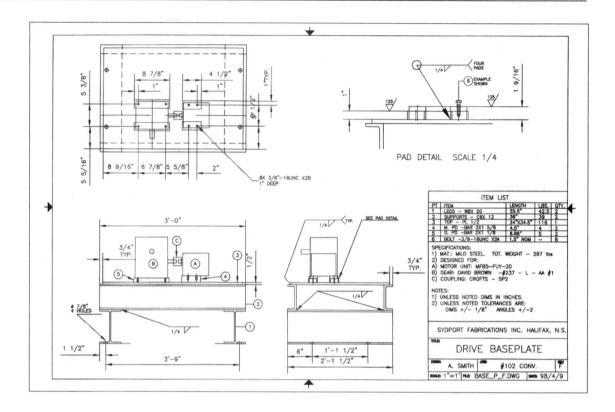

Figure 5.15   Fabrication detail

Assembly Drawings

Most products and designs consist of several parts assembled together. The drawing that describes how they fit together is the **assembly drawing**, sometimes referred to as a **general arrangement** (or **GA**). The purpose of this drawing is two-fold: first to give an overall picture of the completed design; and second, to show the relationship of the various parts that make up the assembly. The parts are listed in a **bill of material** (BOM), **parts list,** or **item list**, usually found on the right-hand side of the assembly drawing sheet. Figure 5.16 shows a parts list on an assembly drawing.

The assembly can be drawn orthographically or pictorially. Orthographic drawings are preferred for engineering assembly, whereas pictorial assemblies are used by people unskilled in drawing-projection methods. Typically, this includes part re-ordering and do-it-yourself kits.

Orthographic assemblies often make use of a section view to show interior parts. All parts are identified by inclined leader lines attached to balloons containing an item or identification number. Parts are listed in the item list along with the quantity per assembly and a description or drawing number (see Figure 5.17).

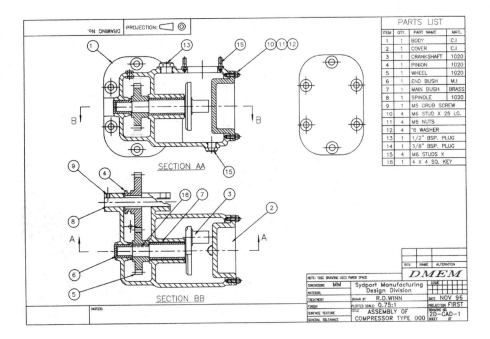

Figure 5.16   Assembly drawing

Dimensions in an assembly drawing are restricted to overall sizes, capacity dimensions, limits or extent of operation, and distances between parts. Other information may be design data and operating instructions. Figure 5.17 shows a dimensioned assembly.

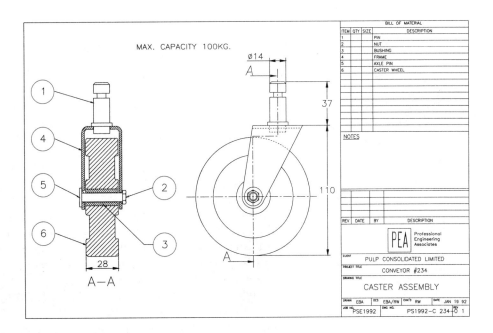

Figure 5.17   Dimensioned assembly

Subassemblies

Subassemblies are small assemblies that are part of a larger design. For example, for an automobile engine, subassemblies would include the oil pump, water pump, transmission, etc. The wheel assembly in Figure 5.17 is a subassembly of a conveying system. Each subassembly may be manufactured and assembled in separate departments and then the units combined to create the final assembly. Figure 5.18 shows a pictorial cut-away subassembly.

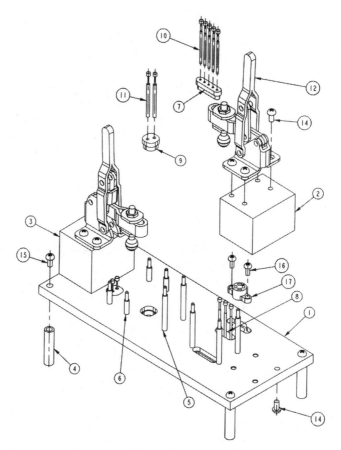

Figure 5.18   Pictorial subassembly [Courtesy: Autodesk]

Building Services

Mechanical engineering drawings used to communicate the design of plumbing, heating, ventilation, and air conditioning systems in commercial buildings, schools, hospitals, etc., are known collectively as **drawings for building services**. The design of the service is generally known as the **system**; e.g., plumbing system, heating system, etc., and the equipment, pipes, ducts, etc., are the **elements** of the system.

A mechanical engineer is responsible for the design of the system, size of the elements within the system, and to some extent the routing and location of the elements. Drawings for building services make extensive use of symbols. It is in the best interest of the mechanical engineer to have a working knowledge of the content and nomenclature of these types of drawings.

Building Drawings

Building drawings are provided by an architect or consulting engineer. They are drawn using orthographic projection techniques but, because of the size, only one view per sheet is common. **Building drawings** comprise plans (similar to a top view), elevations (similar to a side view), and sections. Dimensions are given using the metric system (millimetres); however, the standard system of feet and inches is still commonly used. Mechanical engineers should familiarize themselves with the architect's scale in these cases (see Appendix E).

Most of building drawing applications are of the general arrangement type or an assembly of equipment and connecting pieces. Detail drawings are made to clarify specific features of the system. Detail drawings are enlarged views of equipment installations or components, usually an elevation.

Plumbing

Working with the building architect, the mechanical engineer designs the internal piping systems for the water and gas supply, waste pipes, and drainage. Both orthographic and isometric projection is used to explain the routing and size of pipes. Designs are based on the National Plumbing Code and include washbasins, WCs, showers, etc.

A large facility will require a fire prevention system such as a sprinkler system, which requires a sophisticated valve distribution "station" detail.

Figure 5.19 illustrates a plumbing plan and schematic diagram for an office building. Figure 5.20 shows a sprinkler distribution station for a public administration building.

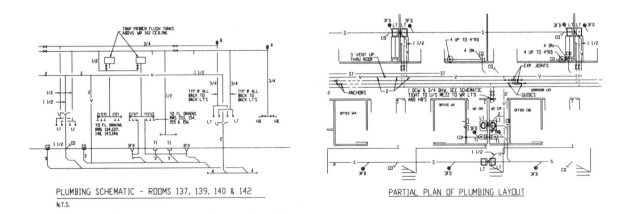

PLUMBING SCHEMATIC - ROOMS 137, 139, 140 & 142
N.T.S.

PARTIAL PLAN OF PLUMBING LAYOUT

Figure 5.19 Plumbing plan and schematic diagram

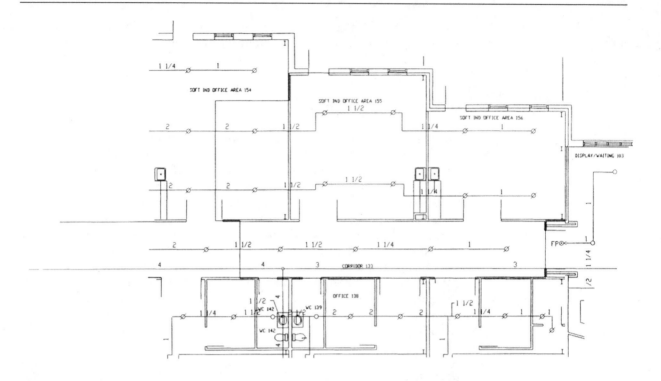

Figure 5.20   Sprinkler layout

Figure 5.21 shows mechanical services details for a heating system (note the detail used on the pump). Figure 5.22 is a plan view of the mechanical room.

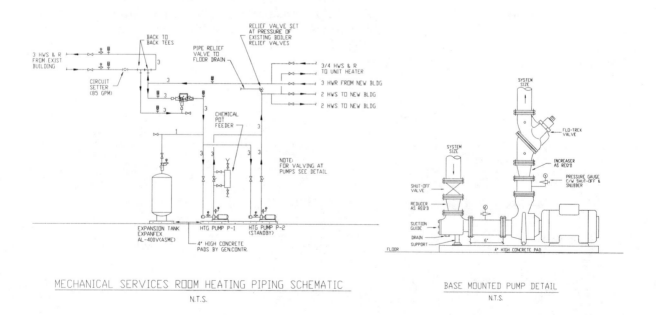

MECHANICAL SERVICES ROOM HEATING PIPING SCHEMATIC
N.T.S.

BASE MOUNTED PUMP DETAIL
N.T.S.

Figure 5.21   Heating equipment drawing

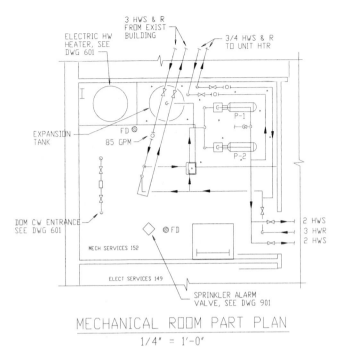

Figure 5.22   Mechanical equipment room plan

Heating, Ventilation, and Air Conditioning (HVAC)

HVAC systems are designed by a mechanical engineer liaising with the building architect or engineering consultant. The purpose of heating and ventilating systems is to help maintain a normal comfort level within the building or to provide a purpose-designed environment (e.g., freezer unit). Ventilation systems remove excess heat, fumes, moisture, pollutants, and odour; and supply fresh air.

HVAC systems consist of mechanical equipment such as air conditioners, heating units, dust collectors, louvers, etc., that are connected to each other using sheet-metal ductwork. Louvers are used to diffuse air into work areas. Detail drawings of individual pieces of ductwork are prepared for specialized or custom applications, making use of developments to create templates for forming the duct.

Generally, drawings for HVAC are similar to those for plumbing; i.e., using architectural plans and elevations. The equipment is shown as an outline and the ductwork may be shown as a single line to scale as illustrated in Figure 5.23. The size of the ductwork is shown as the horizontal dimension followed by the vertical (depth) dimension. Elevation drawings and sections show the height (elevation) of the ducts from a baseline (ground) and the methods of supporting them.

Figure 5.23 shows part of an HVAC plan of an industrial building. Figure 5.24 shows cross-sections (as indicated on the plan) that illustrate the ducting. Note that the principles you learned in earlier chapters of this text are used throughout these drawings. In Chapter 7, we will cover the principles of development used in the fabrication of chute and ductwork.

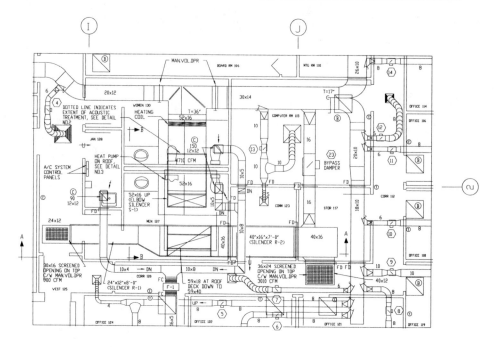

Figure 5.23 HVAC design plan

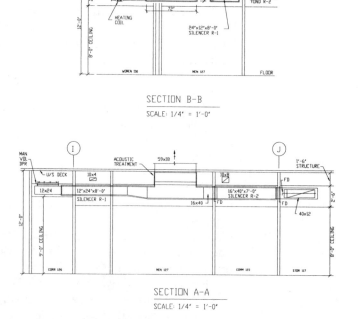

Figure 5.24 HVAC section drawings

Material Handling

Engineering drawings used in materials handling generally make use of orthographic views and sections to describe equipment, storage vessels, conveyors, and associated steel structures. Perhaps overlapping with the structural engineer (see civil engineering drawings), the mechanical engineer is responsible for the design and implementation of such systems. Again, resembling a general arrangement type of drawing, the equipment (e.g., rock crusher, weighing machine, cyclone washer, mine, etc.) is connected by conveyors and chutes. Other connections are to storage bins and loading areas for the material. Supporting structures are also designed and included in the arrangement.

Enlarged detail drawings are made to clarify specific features but generally do not include fabricating information. This is shown on **shop drawings** prepared by trained drafters or technicians from the engineering drawings. If the engineers are working for suppliers of material-handling systems, then they are also responsible for the detail design of the equipment and specifications. A working knowledge of structural steel properties and notations, conveying systems, and other materials-handling equipment is required in this field. Appendix F gives a table of notations used for some common structural steel shapes. When equipment specified for the system is to be supplied by others, detail drawings must be sent to the engineer in order that capacity, size, and connecting details can be verified. Building or site drawings used for the design of the system must always be the latest issue.

Figure 5.25 illustrates a conveyor and chute structure inside a building. The layout shows how the equipment is situated, clearances from other structures, and enough detail for suppliers of equipment to design the requirements.

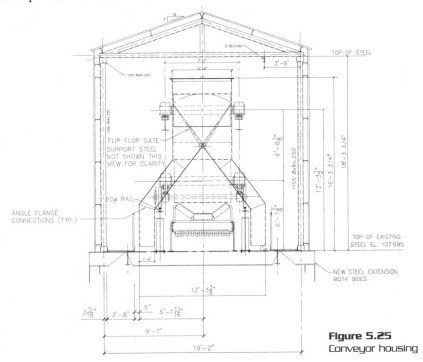

Figure 5.25
Conveyor housing

A general arrangement of materials-handling equipment for use underground is shown in Figure 5.26. Note that the dimensions show the equipment reach and capacity.

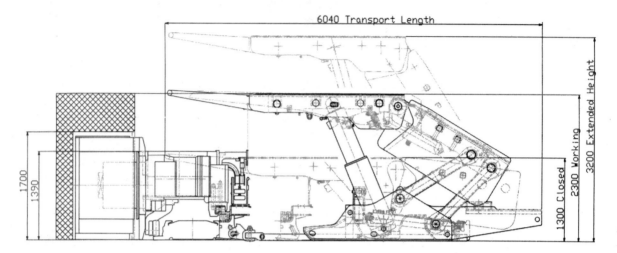

Figure 5.26   General arrangement

Industrial Piping

Industrial piping drawings are used in the design and construction of power stations, aviation fuelling stations, steel plants, coal wash plants, and other applications. Specialized mechanical engineers design systems for transporting fluids such as water, steam, and air through pipes and tubes. Fluids are carried by pipes to and from equipment for processing and storage within industrial facilities. Some cross-over occurs with the chemical engineer as they both design and select equipment for the transportation of fluids.

Pipes are made of steel, iron, plastic, or copper in standard sizes. They are connected to equipment and each other by fittings, that can be welded, screwed, or bolted. Valves are used in piping systems to regulate or stop the flow of a fluid being transported through a pipe.

Engineering drawings for industrial piping show the size and location of the pipes, fittings, valves, and process equipment. To simplify the preparation of working drawings of piping systems, a set of symbols has been developed to represent various pipe fittings and valves. Appendix G lists common pipe sizes and piping symbols.

Piping systems can be so complicated that a 3-D model has to be constructed, either in a workshop or on a computer, to check clearances or interferences of pipes, equipment, support structures, or HVAC equipment. As with all engineering drawings, piping drawings are 2-D representations of the piping model.

The two methods of drawing piping systems, in use, are single-line and double-line drawings (see Figure 5.27a and 5.27b respectively).

Single-line drawings are a diagrammatic representation of the pipes and fittings using symbols. The line drawn represents the centerline of the pipe.

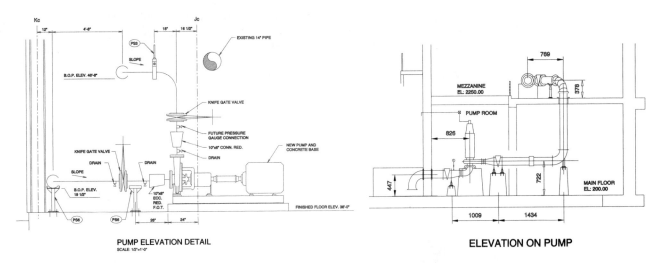

Figure 5.27a Single-line pipe drawing

Figure 5.27b Double-line pipe drawing

Single-line drawings are generally preferred over the more time-consuming double-line ones; however, many companies have their own standards which may, for example, use single-line piping up to 350 (14") diameter pipe and double-line drawings for pipes above that size. The majority of piping drawings are still prepared in feet and inches.

In orthographic drawings, a broken circle (the same diameter as the actual pipe) is used to represent a pipe turning away from the viewer and a solid circle for a pipe turning toward the viewer. In Figure 5.28 an isometric pictorial sketch is converted into a two-view, orthographic detail drawing. Notice how the different pipe locations are shown.

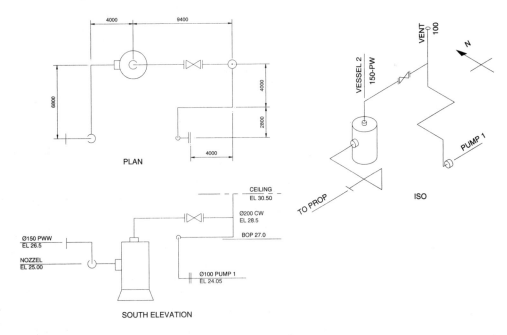

Figure 5.28 Isometric pictorial sketch converted into two-view, orthographic detail drawing

Horizontal dimensions for piping locations are given in the plan view, from center to center of pipe and to the outer face of a connecting flange (see Figure 5.29). Vertical locations are given as elevations from a base (e.g., site zero) or from the flange of an equipment connection (see Figure 5.30) in section views.

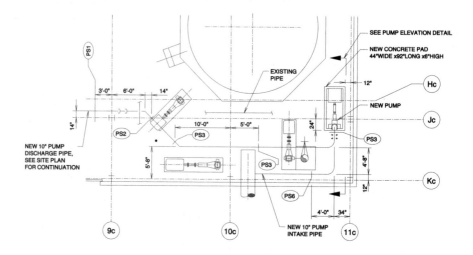

PUMP ARRANGEMENT DETAIL
SCALE: 1/8"=1'-0"

Figure 5.29   Piping design plan

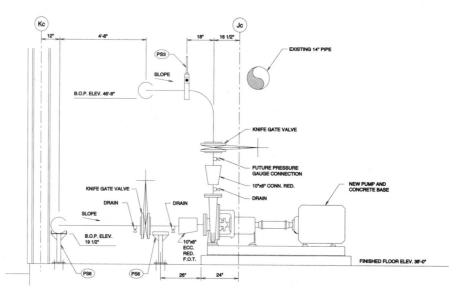

PUMP ELEVATION DETAIL
SCALE: 1/2"=1'-0"

Figure 5.30   Piping design section—From Figure 5.29

Equipment, supports, and vessels are generally shown as a thin line or a phantom outline. The location of these is given on the building layout drawings prepared by the civil engineer.

As industrial plants are generally large, they are divided into areas. The scale of engineering piping drawings is usually 1:50 or 1:20 metric (for inch units, a scale of 3/8" = 1' 0" is preferred). A key plan of the plant is drawn in the upper right-hand corner of the drawing sheet and the area shown on the sheet highlighted in the key plan. Drawing numbers for continuation of piping may be given on "match lines."

The fabrication drawings of industrial piping systems are called **spool drawings**. These can be drawn as isometric or orthographic views. Figure 5.31 is a spool detail of a section of pipe from Figure 5.29.

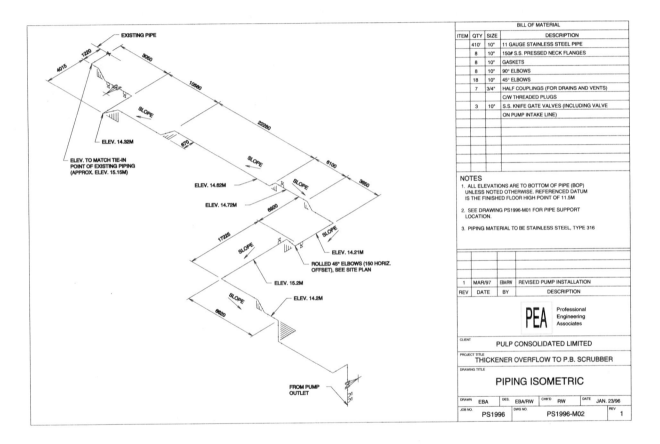

Figure 5.31   Spool detail of a section of pipe

Pipes have to be supported. Generally an engineer is assigned the task of designing piping support systems to withstand the load of the material, the pipe itself, and the shock loading from valving operations. Figures 5.29 and 5.30 showed plan and elevation views with the pipe supports (PS) noted. Further detail drawings will provide enough information for the fabricator to provide shop fabrication details of each support. A typical support detail from Figure 5.29 is shown in Figure 5.32.

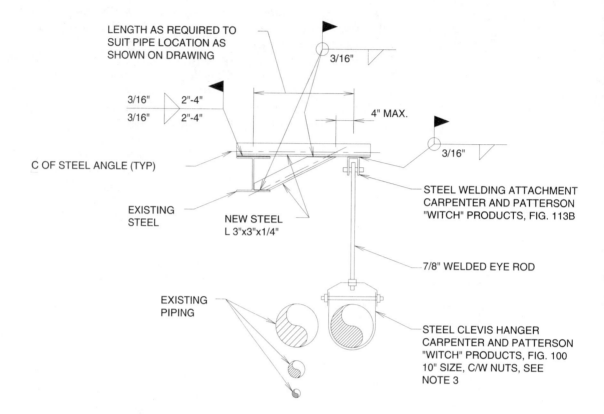

LENGTH AS REQUIRED TO
SUIT PIPE LOCATION AS
SHOWN ON DRAWING

3/16"

3/16"　2"-4"

3/16"　2"-4"

4" MAX.

C OF STEEL ANGLE (TYP)

STEEL WELDING ATTACHMENT
CARPENTER AND PATTERSON
"WITCH" PRODUCTS, FIG. 113B

EXISTING
STEEL

NEW STEEL
L 3"x3"x1/4"

7/8" WELDED EYE ROD

EXISTING
PIPING

STEEL CLEVIS HANGER
CARPENTER AND PATTERSON
"WITCH" PRODUCTS, FIG. 100
10" SIZE, C/W NUTS, SEE
NOTE 3

TYPICAL PIPE SUPPORT DETAIL - PS1
SCALE: NTS

Figure 5.32　Piping supports

Fluid Power

Fluid power is a branch a mechanical engineering dealing with hydraulic
and pneumatic operating systems. A **fluid** is defined as something that
can flow and is able to move and change shape without separating, when
under pressure. Fluid power includes both liquids and gases used exten-
sively in the operation and control of automobile and aircraft systems,
machine tools, heavy-duty equipment, and ships.

Usually the pipes for fluid power are small diameter (bore) tubes of
plastic, nylon, or copper. The engineering drawings generally consist of
schematic diagrams drawn with symbols to illustrate the components of
the system. The symbols are single-line graphics as illustrated in ANSI
T3.28.9-1989 or ISO 1219-1976. As these drawings are schematic, they are
not drawn to scale (see Figure 5.33). The schematic diagram shown
emphasizes the function of the circuit by using symbols to illustrate the
operation of the components. A list of materials and components could be
included as well as specifications for the components.

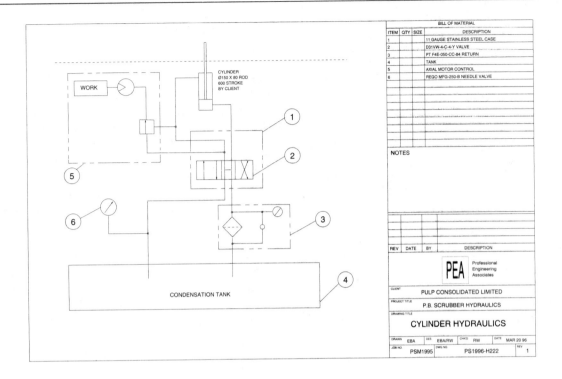

Figure 5.33   Fluid power drawing

Civil Engineering Drawings

Civil engineers also deal with many different types of engineering drawings. The following table lists some of the general areas, that cross over each other in many respects. Much of the work today uses the metric system of measurement; i.e., the millimetre or metre. However, many agencies, especially in the U.S., still use decimal feet as the unit of measure. Engineering drawings using these units are drawn to scale using the civil engineer's scale that uses 1 inch as a base; e.g., 1" = 30' (every inch on paper equals thirty feet on site).

Civil Engineering	Working Drawings Content
Site Preparation and Road Design	Land development and preparation, survey information, GIS/GPS applications, plan and profile, cross-sections, and details
Structural Steel, Concrete, and Wood	Structures: bridges, dams, and buildings. Equipment foundations, tanks and silos
Water and Waste Treatment	Disposal, treatment, and under ground piping

Site Preparation and Road Design

Engineers use their knowledge of surveying, global positioning, and earthwork to design the site for an intended project. The site may encompass drainage, earth removal, and suitable material for foundations to rest on.

Engineering drawings for site preparation start with a topographical plan showing land contours and curves, location of buildings or structures, roads, water and utility lines, and the proposed project boundaries or traverse. **Contours** are curved lines that represent a certain elevation above a known point, usually sea-level. The curves are drawn closer together to represent a steeper gradient or change in elevation. The elevation is indicated on the contour line. Contour lines are discussed in more detail in Chapter 6.

The location of the site boundaries and proposed structures are given by distances from a known point together with the angle. The angle can be given as either an **azimuth** (0-360 degrees) or a **bearing** (measured from a quadrant; e.g., N45°W is the same as 315° as an azimuth). You will learn more about bearings and azimuths in Chapter 6.

A grid system is often used giving distances in a north or east direction. Figures 5.34 and 5.35 are examples of site drawings.

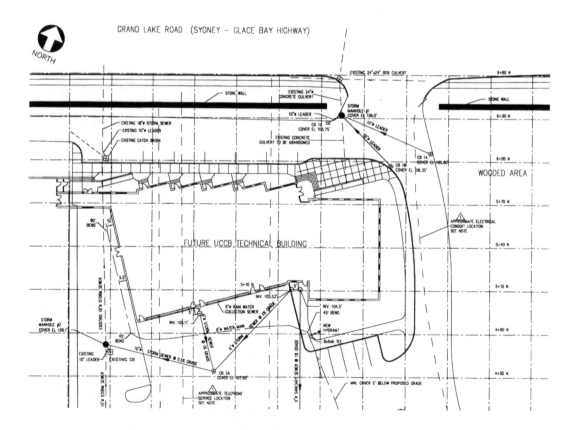

Figure 5.34   Site plan for a new building

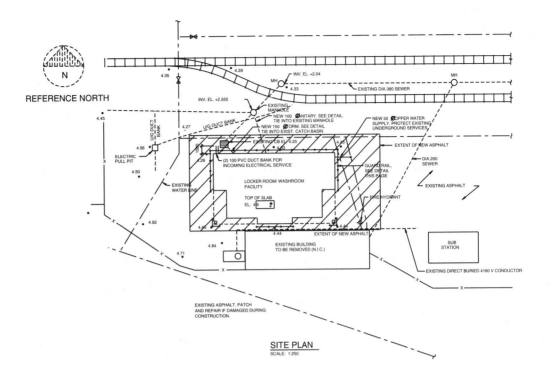

Figure 5.35   Site plan for renovations

Highway design drawings start with a **key plan drawing** of the entire area involved in the project. This is divided into sections so that the drawing scale is manageable. Each section is drawn as a topographical plan drawing, showing the land layout. This includes contour lines expressing the altitude of the land and a grid system giving distances in a north direction and east direction. Features such as rivers, lakes, buildings, and railroads are included on the plan, which is drawn at a scale convenient for the area.

The proposed highway is plotted onto the topographical plan, using survey information and design criteria. For identification purposes, the highway is divided into "stations" along the centerline. Typical scales for plan drawings are 1:1000 or 1:500 metric (1" = 100' or 1" = 50' standard).

Plan and Profile Drawings

Below the plan, a full-section drawing is created—called a **profile drawing**. Generally the profile is taken at the centerline of the highway and is drawn to an exaggerated scale to describe differences in elevation more clearly. Profile drawing scales are typically ten times that of the plan; i.e., 1:100 or 1:50 (1" = 10' or 1" = 5'). The surface of the natural ground is labelled "existing grade" and is shown as plotted from survey data. The centerline of the proposed highway sometimes cuts into the existing ground or requires filling to raise it to the design elevations.

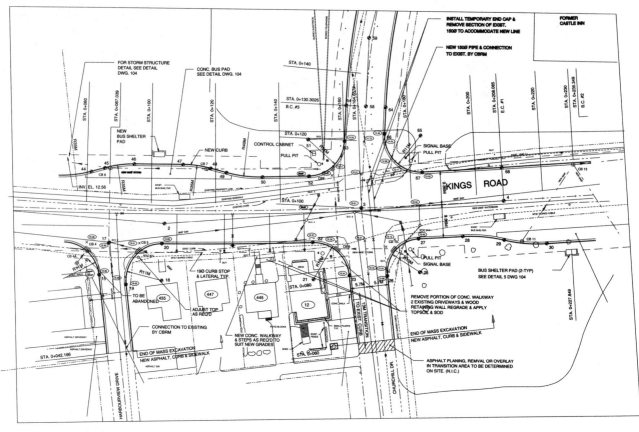

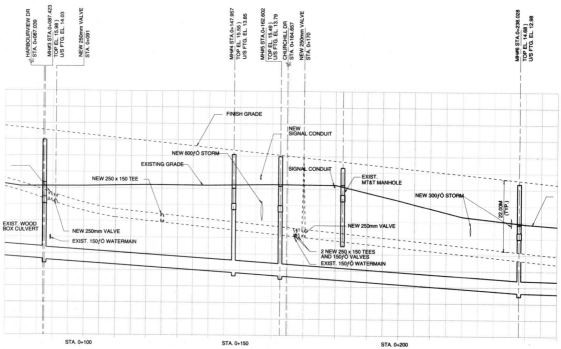

-1.00% GRADE

KINGS ROAD CENTERLINE PROFILE
SCALE 1:400 HORIZONTAL
1:40 VERTICAL

Figure 5.36 Plan and profile drawing of proposed highway interchange

The material to be cut or filled is shown on the profile drawing. The material that has to be removed or "borrowed" can be a decisive factor in the cost of the project, in relation to the number of transport operations, and hence the time and money required to prepare the site. Figure 5.36 is a plan and profile drawing of a proposed highway.

Cross-section Drawings

As a profile drawing is taken along the longitudinal centerline of the road or site, **cross-section drawings** show the conditions at various points perpendicular to the centerline. In order to accurately determine the amounts of material to be removed or added, cross-sections at each station line are created. The roadway or structure is superimposed on each cross-section, which is again drawn at a larger scale than the plan to show detail. A typical road cross-section is shown in Figure 5.37.

In the case of highway construction, the cross-section drawings include details such as depth of material (subgrade), slope of the road for drainage, culverts or drains to remove water, super-elevation details for curves in the road, and often the location of underground pipes and conduits, curb details, and lighting.

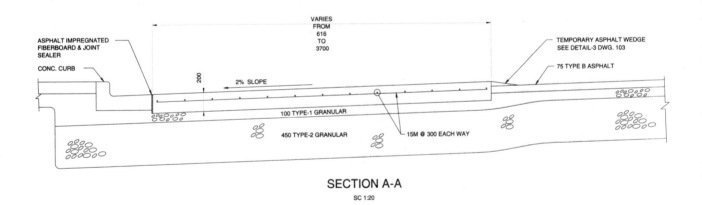

SECTION A-A
SC 1:20

Figure 5.37   Road cross-section

Structural Steel, Concrete, and Wood

Structural engineering is generally associated with large steel, wood, or concrete structures and buildings. A structural engineer works with designers and architects to engineer the structural components required to determine the sizes of building supporting members, foundations, and geotechnical requirements. Although under the same heading, structural, steel, and concrete engineering drawings differ in the amount of content. Wood structures can be framed buildings or intricate, ornate, arched structures for public buildings.

Structural drawings generally fall into two categories: engineering design drawings and shop (manufacturing) drawings.

Engineering design drawings are set of working drawings that completely describe the project. Drawings include a site location plan, design layout plans and elevations of the locations of structural members, foundation layouts and concrete details, building elevation views, and typical wall sections. Figures 5.38 through 5.40 show several design working drawings for a small building.

The engineer must become familiar with the terminology used on structural design drawings. The method of indicating structural steel members is given in ANSI and CSA standards (see following example). Structural members that are repeated on a plan are noted with "Do" meaning ditto (see Figure 5.41).

The design drawings for steel structures are primarily concerned with clearly showing the structural elements' size and location. Details are left to the fabricator who prepares shop drawings.

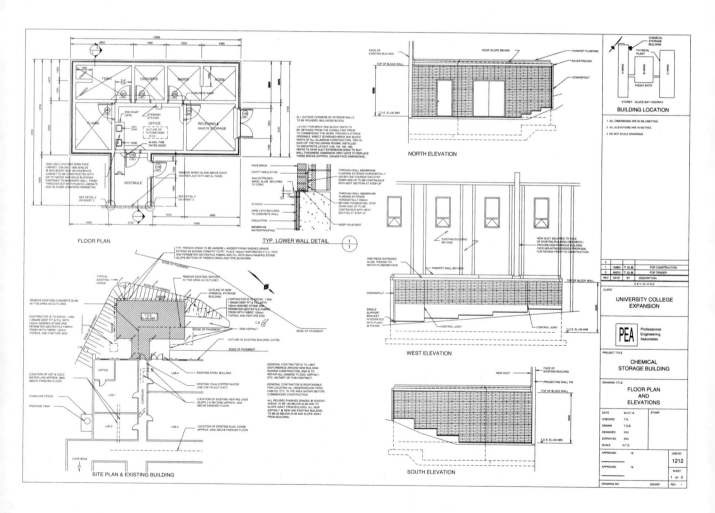

Figure 5.38   Design working drawings—Sheet 1 of 3

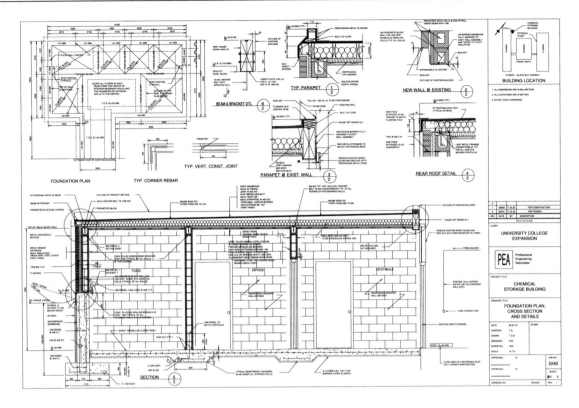

Figure 5.39   Design working drawings—Sheet 2 of 3

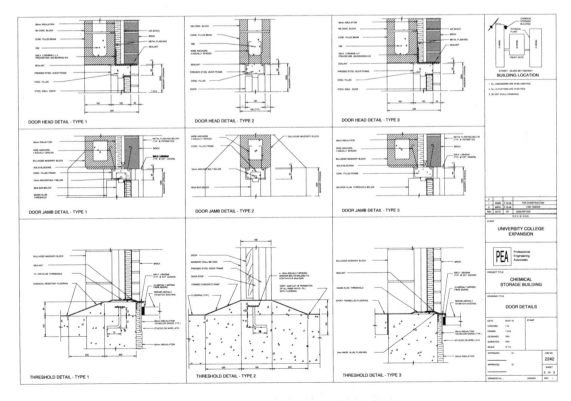

Figure 5.40   Design working drawings—Sheet 3 of 3

Shop (manufacturing) drawings for steel structures are generally made by the steel fabrication supplier.

Structural drawings illustrate the details and dimensions of all pieces of the structure and how they are to be fabricated (clearances, weld information, etc.). Each piece is identified by a marking system and listed in a bill of material (BOM). The fabricator provides an **erection drawing** or an **assembly drawing** showing the parts in situ (which is similar to the design drawing). The following figures show design drawings: Figure 5.41 shows a structural roof plan and Figure 5.42 shows a section from the plan.

Drawings for concrete construction include all of the information necessary to install the concrete structure: size and placement of concrete, ground requirements, reinforcing bar information, and connection details to steel or wood structures. Bar sizes are indicated as follows:

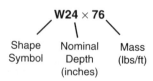

20M @ 200 E.W
means 20 millimetre diameter bar spaced at 200 millimetres in both horizontal directions (Each Way).

Although concrete construction drawings follow standard engineering projections, they often appear cluttered. Figure 5.43 shows the concrete details for a steel-framed building.

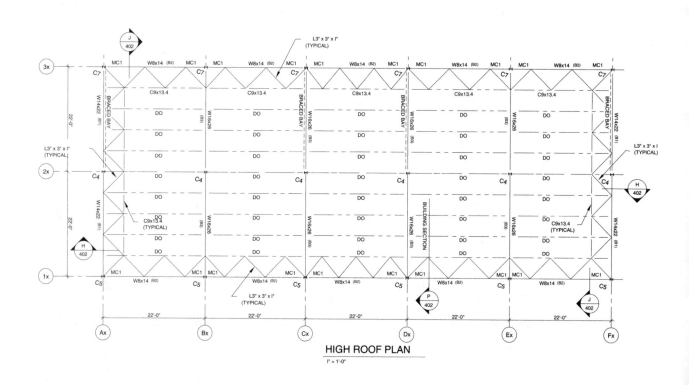

Figure 5.41 Structural roof plan

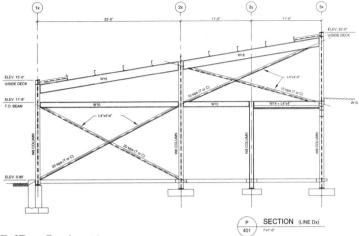

Figure 5.42   Steel section

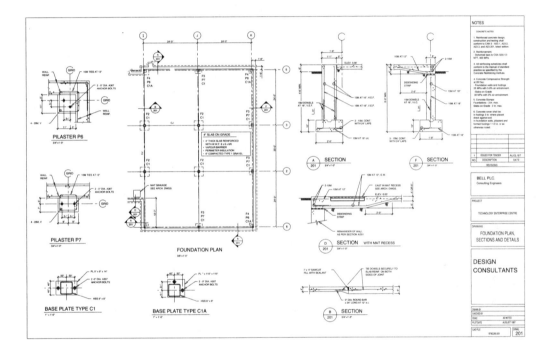

Figure 5.43   Concrete details

Water and Waste Treatment

Civil engineering incorporates the design of fresh, waste, or sanitary water transport; and underground drainage. These drawings are termed **underground services** for the civil project.

The pipe and drainage information is shown on a plan and profile drawing, as discussed in the previous section. In addition, detail drawings of disposal beds and underground fixtures may be included. Figure 5.44 shows a plan and profile drawing of a proposed sewer and water line. Figure 5.45 shows a typical detail sheet for underground fixtures.

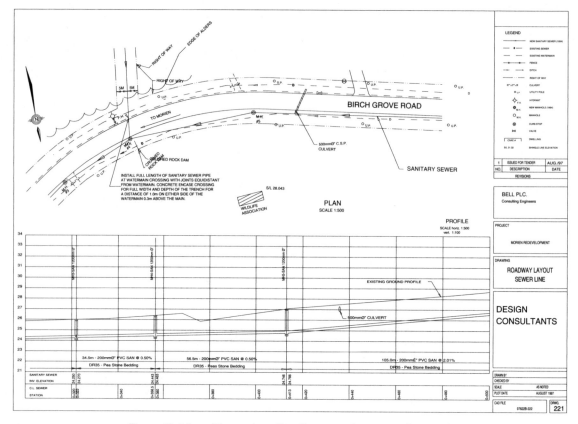

Figure 5.44 Plan and profile of proposed sewer and water line

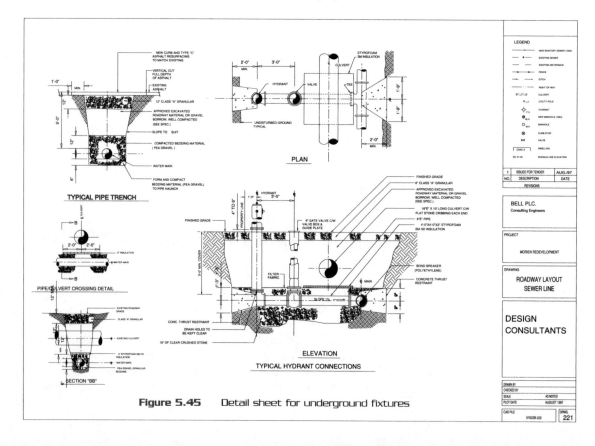

Figure 5.45 Detail sheet for underground fixtures

Chemical Engineering Drawings

Chemical engineering involves equipment and processes for the transportation and manufacture of chemicals. Many of the drawings involved with chemical engineering are similar to those encountered in mechanical engineering; i.e., piping, vessel design, and material-handling equipment.

The chemical engineer is more concerned with the process involved and the capabilities of the equipment. Therefore, the drawings initially used are schematic-type diagrams indicating the flow of material and the process steps. Because of the nature of industrial competition, or for security reasons, most firms use a code to name chemicals or processes. Alternatively, the chemicals and processes may be left off the drawing entirely.

Process and Instrumentation Diagrams

The **process and instrumentation diagram** (P&ID) is the chemical engineer's road map of the system. It shows all of the major equipment and information that is relevant to the process in a schematic form; for example, equipment names and numbers, the sizes of pipes and line numbers, valves, capacities, pressure ratings, instrumentation, and any relevant data tables.

The drawing is not scaled. It is used to draw the design piping drawings in conjunction with the plant drawings. The P&ID makes use of standard symbols such as those found in ISO 3511 and should be laid out consistently and evenly. A partial P&ID is illustrated in Figure 5.46.

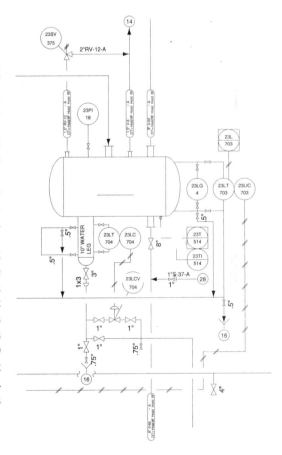

Figure 5.46 P&ID

Electrical and Electronic Engineering Drawings

Electrical engineering deals mainly with power generation and distribution and is used in everything from industrial plants to computers. Control of electric motors, heating and lighting systems, and industrial processes all require electrical engineering drawings.

Electronic engineering is concerned with using power in much smaller quantities for the transfer of electrons to make things happen. Printed circuits, computers, audio/visual equipment, satellites, and instrumentation are all part of the electronic engineer's domain.

The types of drawings used by electrical/electronics engineers vary according to the particular field they are in. The majority of the drawings are schematic-type diagrams; i.e., using symbols to represent components; not drawn to scale. Their main purpose is to communicate how electrical/electronic devices work together. The major types of diagrams are listed below.

- Schematic diagrams
- Connection and wiring diagrams
- Printed circuit board diagrams

Schematic Diagrams

Schematic diagrams are very common in electrical/electronic engineering. They show the functional relationship and connection of components used in an electrical/electronic circuit. Schematic diagrams do not show the physical size of the components nor the circuit. Often components are rotated or inverted from their actual orientation to improve the drawing layout. In electrical engineering, single-line diagrams that show power distribution to equipment throughout a project are common. They are simplified schematic diagrams that show only the major components and the power feed to them. Figure 5.47 is an example of a single-line electrical schematic diagram used in the construction of a public health facility.

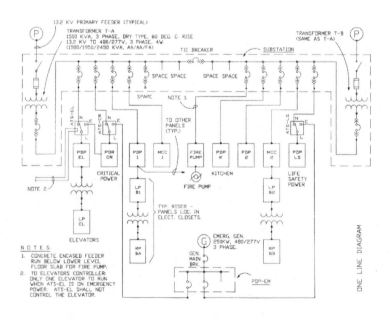

Figure 5.47 Single-line electrical schematic diagram

Electronic schematic diagrams use a component numbering system that can generate a bill of material list or be converted into a net list for use with other software. Typically, an electronic schematic diagram is the circuit design that is translated into a printed circuit board or "chip" layout for production. Figure 5.48 illustrates a partial schematic diagram for a communications satellite.

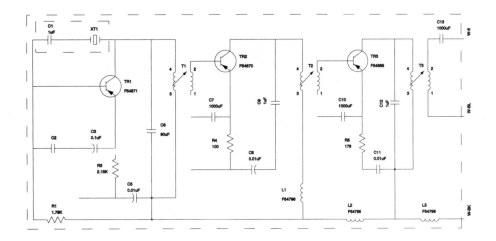

Figure 5.48   Partial schematic diagram for a communications satellite

Connection and Wiring Diagrams

Connection diagrams are used to show how components in an electrical assembly are connected. The types of connection diagrams include **point-to-point, highway, baseline,** and **harness diagrams**. Figure 5.49 shows a highway connection diagram. Each terminal of each component is numbered. Connecting wires often use a color-coding scheme to lessen the chance of error during assembly. The components are drawn to represent the shape of the component and are placed in relative positions to each other, unlike the symbols used in schematic diagrams.

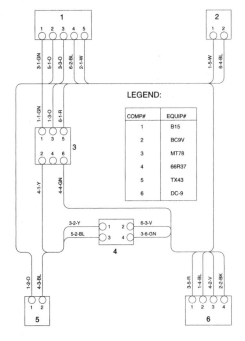

Figure 5.49   Highway diagram

The term **wiring diagram** is used in industrial and commercial wiring applications. One common method is to show the connections on a *ladder*. The **ladder diagram method** is used in motor control circuits, alarm circuits, and programmable logic controllers. A typical motor control circuit is shown in Figure 5.50. Note that the power supply is shown at the top of the diagram, then a transformer to reduce the voltage to the control circuit is shown below. Each line is usually numbered.

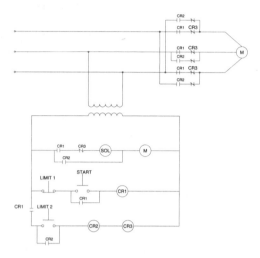

Figure 5.50   Ladder diagram

Again, the symbols are described in the appropriate standards.

Printed Circuit Board (PCB) Diagrams

Most electronic devices use a circuit board to mount components and wiring connections between components. The board is made of a laminated insulating material (e.g., fibreglass) and the connections made by a thin layer of conductive material (e.g., copper) etched into the board. These copper **traces** act as wires to connect the components as designed in the schematic diagram. The traces are generally 0.2 to 0.5 mm wide and 0.05 mm thick, depending on the current carried and the number of components on the board. The components can be resistors, capacitors, transistors, etc. and integrated circuits (IC) or chips. **Chips** are miniaturized printed circuits, for very low current, made of silicon.

The components can be mounted on one or both sides of the board with holes that are plated through to carry current from one side to the other. Traces on both sides are noted by using different colours or line types (e.g., dashed for the other side).

Drawings to manufacture PCBs are done in different stages, and are more mechanical in nature than electronic. The traces and component layout are done on an **artwork drawing**. A **master drawing** shows the board geometry, connector pattern, mounting holes for the components, and manufacturing dimensions. The board is designed to fit in its appropriate space within the equipment. The components have to be mounted in pre-drilled holes located with precision on a **drilling drawing**. An assembly drawing is used to show the components mounted in position. Figure 5.51 shows a PCB master drawing, used to develop the working drawings for production.

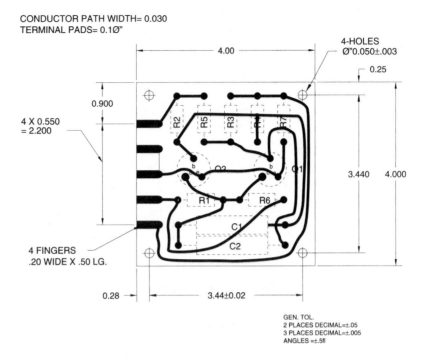

CONDUCTOR PATH WIDTH= 0.030
TERMINAL PADS= 0.1Ø"

Figure 5.50 Printed circuit board master drawing

Standards and Specifications

Standards

Almost all engineering drawings are governed by standards of one form or another. **Standards** refer to a set of symbols or methods of illustration that have been accepted throughout the industry and adopted into a national or international document. Applying standards to engineering drawings saves time and money. Time is saved because the standards already exist and are recognized throughout the industry. There is no need to create or re-design parts and components that have been standardized.

Threaded fasteners, springs, gears, and many holding devices are examples of standardized mechanical parts. For economic reasons, standard shapes, sizes, and tolerances for steel and wood products are to be used whenever possible. Electronic symbols, circuitry, and power supplies are all available as standardized symbols.

The way items are shown on an engineering drawing should be in accordance with the applicable standards to prevent misunderstanding. The following table lists the various standards pertaining to engineering drawings in the disciplines we have discussed.

General presentation	ANSI Y14.100	ISO 128
Dimensioning	ANSI Y14.5	ISO 129
Tolerancing	ANSI Y14.5	ISO 406 / 1101
Welds	AWS A2.4	ISO 2553
Process symbols	ANSI Y32.11	ISO 3511
HVAC	ANSI Y32.2.4	
Piping symbols	ANSI Z32	ISO 6412
Building		ISO 2594
Civil engineering		ISO 3766/7084
Electrical symbols	ANSI Y32.2d	

Specifications

Specifications include any standards the designer wants incorporated or adhered to; for example, steel specifications may relate to an ASTM (American Society for Testing Materials) standard for the composition and abilities of steel.

Drawings must convey the designer's intent as far as the geometric and dimensional characteristics are concerned. Specifications are written so that the manufacturer will make or build the objects on the drawing in accordance with the designer's requirements for material, finish, standards to be adhered to, and any particular methodology to be followed.

Specifications are sometimes added to the drawings, in note form, on the right-hand side. On large projects, the specifications are usually produced as a separate document, issued with the drawings. If the information given in the specifications differs from that given in the dimensional content of the drawing, generally, the drawing overrules.

Writing specifications is often the job of a senior engineer, as the wording has legal implications; however, many companies use juniors to prepare the "specs" and add them to the drawings or related documents.

For civil engineering projects in Canada, a National Master Specification has been written. It can be used by editing and adding in the appropriate wording.

Now that you know a little more about engineering drawings, try the following problems.

Problems

Problems for this chapter are at the discretion of your instructor, as each of you will eventually complete your engineering instruction in a specific discipline area. Your instructor may ask you to prepare working drawings based on the sample drawings in this or other chapters.

Complete all problems on a title sheet designed for your course, and lay them out in accordance with engineering graphics' practices.

1. Create sufficient views to fully describe the part shown below. Include an auxiliary view. You may use partial views if sanctioned by your instructor. Add dimensions to the views according to standard practices. Tolerances may be added at the discretion of your instructor.

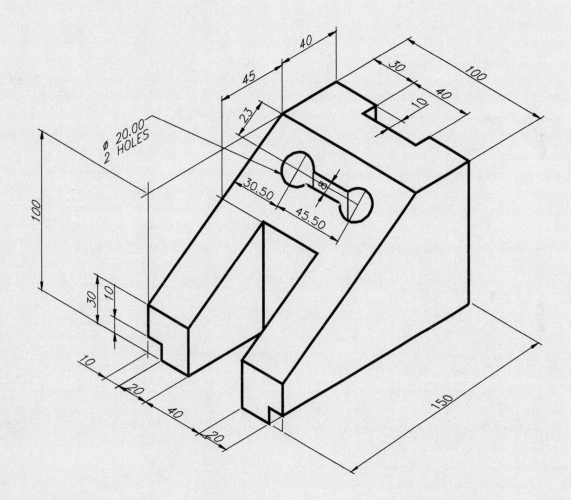

Figure 5.51

2. From the details given below, create an assembly drawing. The axle goes through the wheel with an axle support on each end, facing out. The axle supports bolt onto the top plate. The drawing should have at least one full cross-section view and one regular view. If directed by your instructor, include a bill of materials (parts list) and details of bolts, nuts, etc.

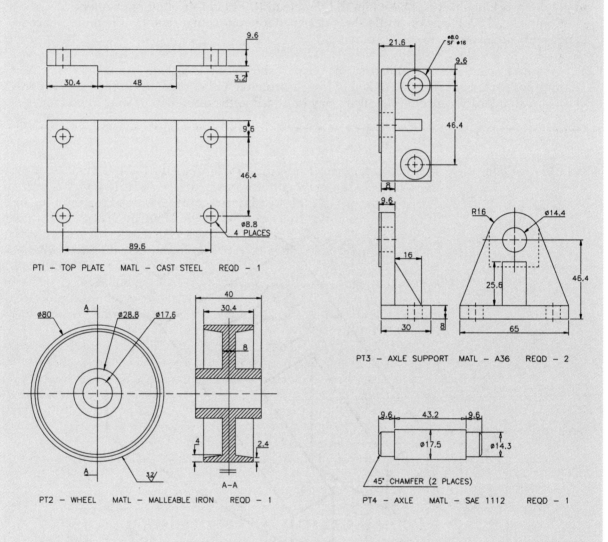

Figure 5.52

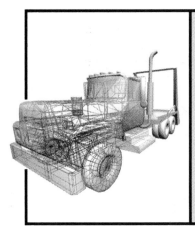

Chapter 6
Visualization

As noted in Chapter 1, although we live in a three-dimensional world, we must represent it on two-dimensional paper or on a computer screen.

Isometric and perspective drawings are two ways of representing three dimensions on a two-dimensional surface. Mathematically, three dimensions can be represented by Cartesian coordinates, x, y, z, referenced to some origin and a set of axes. Position on the face of the earth can be defined by latitude, longitude, and elevation, above or below some datum, usually sea level. Latitude, longitude, and elevation are used for surveying and navigating. We also need a method of finding qualitative information on three-dimensional space on a two-dimensional surface. This is called **descriptive geometry**.

What an object looks like depends on its orientation and the viewpoint from which it is seen. To see how close one object is from another, you must choose the correct viewpoint. The procedure is similar whether you use paper or a computer. With paper, you create views (called **auxiliary views**) in addition to the regular front, top, and side views. With a computer, you set the point from which the objects are seen. The two methods are covered in separate sections of this chapter.

Let's start off by looking at some common visualization terminology.

Visualization Terminology

Three common visualization terms are point, line, and plane.

Point

The **point** is the basic building block for an object in space. It has location but no dimensions. It can be located by assigning Cartesian coordinates, x, y, z, from an origin or by a distance from an arbitrary datum.

A point is represented on a two-dimensional surface by a small cross or by the intersection of two lines. Figure 6.1 shows front, top, and side views of a point, A.

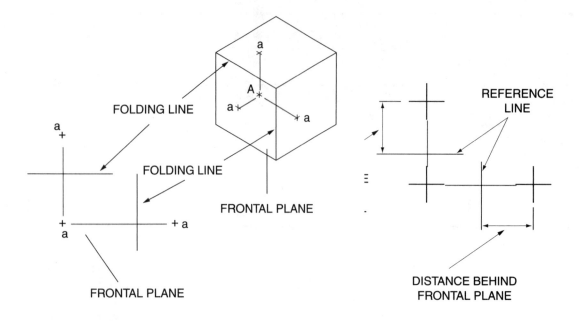

Figure 6.1 Three views of a point

Figure 6.2 Reference distance from the picture plane

The folding lines in the orthographic views correspond to the folds between planes of an enclosing box. The location of the point is specified with reference to these folding lines. Since (with third angle projection) the picture plane is always between the viewpoint and the point, the point is always behind the plane. The distance from the picture plane is used as a reference distance and the datum is the folding line. Figure 6.2 shows where this reference distance appears on orthographic views.

The distance behind a plane, the frontal plane in this example, is seen in any and all views adjacent to the frontal plane. If the distance from the folding line is known in one adjacent view, it can be used to locate the point in any other adjacent view. The plan view and the right-side view are both adjacent to the front view. This same distance is used to locate the point in a left-side view.

Line

A straight line is made up of two points. Figure 6.3 shows three views of a line, AB. Note: The frontal plane is designated plane 1.

The ends of the line are identified by crosses.

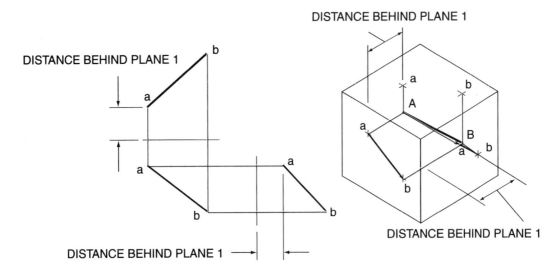

DISTANCE BEHIND PLANE 1

DISTANCE BEHIND PLANE 1

DISTANCE BEHIND PLANE 1

DISTANCE BEHIND PLANE 1

Figure 6.3   Three views of a line

Plane

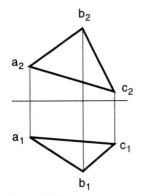

Figure 6.4
Two views of a plane

A **plane** is a flat surface such that a straight line joining any two points lies in the plane. The orientation, but not necessarily the limits, is defined by a minimum of three points. These points can be anywhere in the plane; they do not have to be at the edges (although they often are). Figure 6.4 shows two views—front and top—of plane ABC, made up of three points and three lines.

A subscript indicates the plane on which the point is projected; e.g., a_1, b_1, c_1. As mentioned previously, the frontal plane is arbitrarily designated as plane 1. When there are several auxiliary planes, they should be identified with numbers. Most problems, however, do not require more than two auxiliary planes.

Let's look at the properties of lines and planes.

Properties of Lines and Planes

Let's start off by looking at the properties of lines.

Properties of Lines

Four pieces of information are needed to define a line:

- the location of one point on the line (which we have already discussed),
- direction,
- the angle the line makes with the horizontal (the **slope**),
- length.

Direction

Direction, given as a compass reading, is seen only in the horizontal plane (the plan view) because when you read a compass, you do not hold it in a vertical position. There are two ways to indicate a direction: bearing and azimuth. A **bearing** is measured from either north or south (see Figure 6.5).

The angle is always less than 90°. A bearing of N 90°W would be given as west.

An arrow indicating north must be shown in the plan view. The "north arrow" usually points to the top of the page.

An **azimuth** is a compass direction measured from north and stated as an angle from 0 to 360°. Figure 6.6 shows how an azimuth is measured.

Since an azimuth is always measured from north, there is no need to specify a compass direction, but sometimes a compass direction is given.

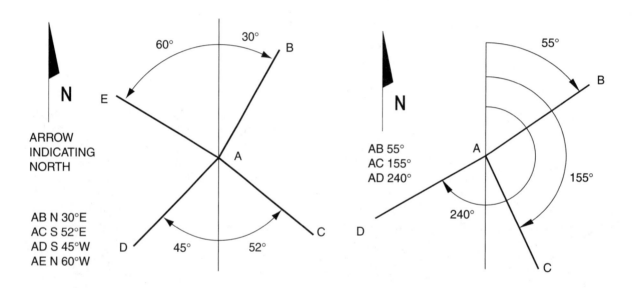

Figure 6.5 The bearing of a line   **Figure 6.6** An azimuth of a line

Slope

A slope indicates whether a line rises or falls from one end. You have probably used "rise over run" to describe slope. **Slope** is defined as the angle between a line and the horizontal plane (see Figure 6.7). It can be seen and measured only in an elevation view where the line appears as true length. In order to measure slope, you must be able to find the true length of a line. If slope is measured by "rise/run," both rise and run must be seen in true length.

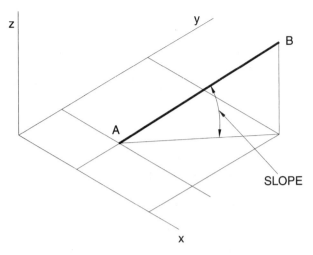

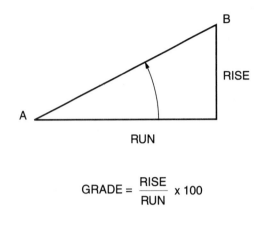

Figure 6.7 The slope of a line

Figure 6.8 Slope expressed as grade

Civil engineering applications give slope as a grade. **Grade** is the tangent of the slope expressed as a percentage (see Figure 6.8). A 100% grade is equivalent to a slope of 45°.

The run does not have to be 100, but calculation is much simpler using 100. You can use any multiple of 10, but a run of 100 gives more accurate results.

Length of a Line

A line can be viewed from any direction. It will appear as a different length depending on how it is viewed. Length can be measured, but only if the line is first seen in true length. A line can be seen in true length only if you look in a direction perpendicular to it. When viewed from any other direction, it will appear as some length between these limiting cases. We must determine how to get a view that shows true length. We will see how to find the other limiting case, a point view, later.

Finding the True Length of a Line

Figure 6.9 shows front and plan views of a line, **AB**. It is not true length in either view, as we will see later. To find the true length of line **AB**:
1. The line must be viewed from a perpendicular direction to see true length. Choose a view that satisfies this condition. A side view (either left or right) satisfies this condition for this line.
2. Draw a folding line between the frontal plane and the right-side plane so that the end points can be projected onto the right-side plane. The location of the folding line is arbitrary as long as it goes in the correct direction.
3. Project points A and B into the right-side view. Figure 6.10 shows how the end points are projected. The distance from the folding line in the plan view is used to locate the end points in the side view.

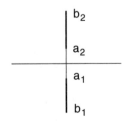

Figure 6.9
Two views of a line

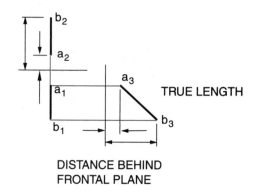

DISTANCE BEHIND
FRONTAL PLANE

TRUE LENGTH

DISTANCE BEHIND
FRONTAL PLANE

Figure 6.10 Locating points in a third view

There is no need to measure this distance. It can be transferred with dividers. The true length of AB is seen and can be measured in the right-side view.

This example is a special case in which the true length appears in one of the principal orthographic views. This is because the line is parallel to one of the principal planes. The isometric shows the line is parallel to the sides of the enclosing box before it is unfolded.

The same process is used to find the true length of the line shown in Figure 6.11. This line is not parallel to any principal plane.

1. Locate an auxiliary plane such that the line of sight is perpendicular to AB. The folding line representing this auxiliary plane is located parallel to line AB. (The auxiliary view can be taken from the plan view or the front view. The plan view is used here.) Since lines of sight are always perpendicular to picture planes, they will be perpendicular to AB (see Figure 6.12).

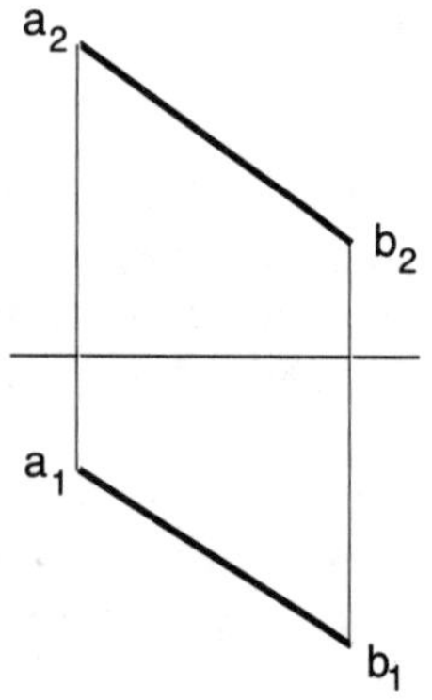

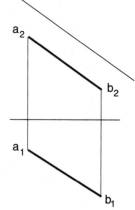

FOLDING LINE PARALLEL TO AB

Figure 6.11 Front and plan views of a line

Figure 6.12 Location of auxiliary plane to show true length

Figure 6.13 is an isometric drawing showing the picture plane for the auxiliary view relative to AB.

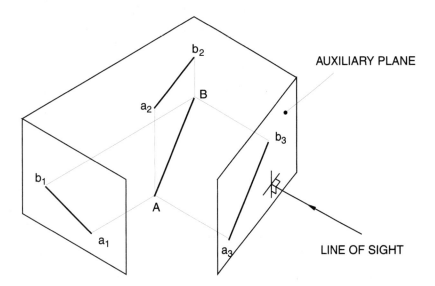

Figure 6.13   Locating an auxiliary plane to find true length

2. Project the end points onto the auxiliary plane (called plane 3). Distances from the folding line on the auxiliary plane are taken from the frontal plane, as shown in Figure 6.14. The true length (TL) can be measured in plane 3.

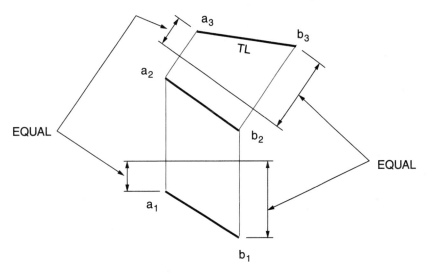

Figure 6.14   Projecting points onto an auxiliary plane

An auxiliary plane was taken off the plan view in this example but it could have been taken off the front view. In some cases, an auxiliary view must be taken off a particular view, but in this example it does not matter. If possible, locate the auxiliary plane so that it does not run off the paper.

Point View of a Line

A point view is required to solve some line problems.

In order to see a line as a point view, you must "look along" the line. The line must be in true length before you can do this, so you must find a view showing the line in true length.

Finding the Point View of a Line

As you may recall, we found the true length of the line in Figure 6.14. This figure is repeated as Figure 6.15.

1. After finding the true length, locate an auxiliary plane (plane 4) so as to "look along" the line. The folding line is drawn perpendicular to the true length. Either end can be used.
2. The point view is positioned by taking distances from the plane adjacent to the true length view (the plan view in this example). The point view is seen in Figure 6.16.

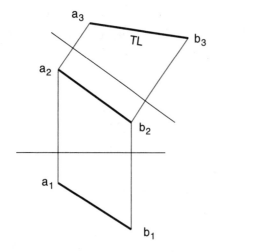

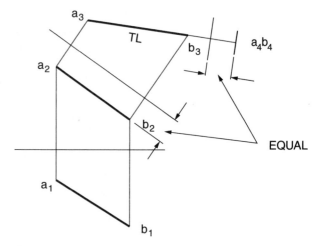

Figure 6.15   View showing a true length (Figure 6.14 repeated)

Figure 6.16   Finding a point view

The distance must be the same for each end of the line. If they are not the same, there is an error.

Slope of a Line

Slope can only be seen in an elevation view showing the line in true length. This is so that the rise and run are seen correctly.

Finding the Slope of a Line

Figure 6.17 shows two views of line AB. The slope is to be drawn and measured. To do this:

1. Place an auxiliary elevation view so that AB appears as true length. The folding line is drawn parallel to AB in the plan view. It can be located on either side of the line.
2. Draw the line in the auxiliary view. The distances from the folding line are taken from the front view (adjacent to the plan view). Figure 6.18 shows the true length.
3. Slope is the angle between the line and a horizontal line (in the auxiliary plane). Any line parallel to the folding line in this auxiliary view is a horizontal line. (The folding line between the plan and auxiliary view is also a horizontal line.) Figure 6.19 shows the slope.

A line has a positive slope if it gets closer to the folding line and a negative slope if it gets further from the folding line when going from one end to the other. The slope from A to B is positive and from B to A, negative. The slope in Figure 6.20 is shown as a grade.

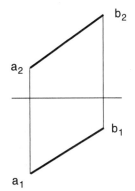

Figure 6.17
Two views of a line

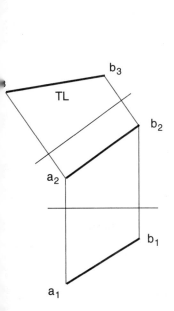

Figure 6.18
Elevation view showing true length

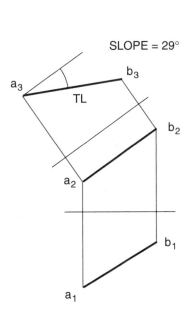

Figure 6.19
Measuring the slope of the line

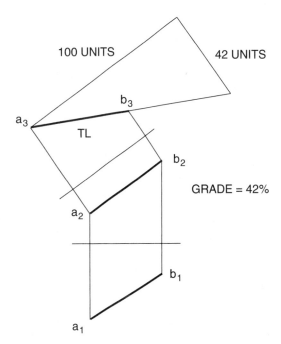

Figure 6.20
Slope expressed as a grade

The grade can be read directly from a scale by laying out a horizontal run of 100 units (any convenient scale can be used) and measuring the vertical rise to the same scale (42 units in our example).

As grade = rise/run × 100, in our example the grade is 42/100 × 100 = 42%.

To help you understand the practical application of your knowledge, take a few minutes to look at the following examples.

Example 6.1

A wire rope supporting a broadcast antenna is anchored to the ground at elevation 400 m and to the antenna 1 m below the top. The antenna is 50 m tall and the base is at an elevation of 390 m. The wire is anchored on the ground 14 m east, 20 m north of the mast. Determine the length and slope of the wire. The scale is 1:500.

What is known?

- Location of the anchor, relative to the mast
- Elevation of the anchor
- Elevation of the attachment near the top of the mast
- Elevation of the base of the mast and the height of the mast.

1. Determine what view is necessary to find the required information—true length and slope. Both can be measured on an elevation view showing the true length of the line.
2. Lay out a plan view showing the location of the anchor and the mast. Figure 6.21 shows the plan view, drawn to scale.

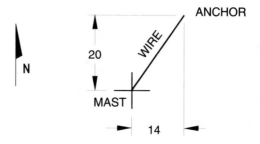

Figure 6.21   Locating points in the plan view

3. Lay out an elevation view to show the elevations of the anchor, mast, and wire. Draw an elevation view that will show the true length and slope of the wire. (This is not the front view.) The folding line is located parallel to the wire to show true length in the auxiliary elevation view. The ground elevation 400 m can be located anywhere. Figure 6.22 shows the elevations and plan view.

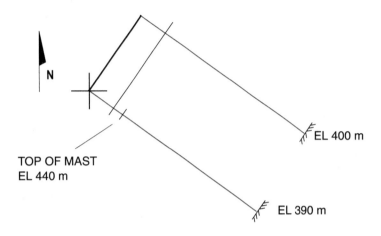

Figure 6.22   Elevations and plan view

4. Draw the wire from the anchor to a point 1 m below the top of the mast. Measure the length and slope of the wire. Figure 6.23 shows the length and slope.

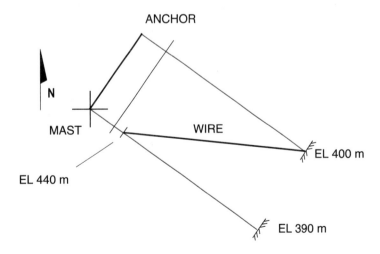

Figure 6.23   Length and slope of guy wire

Example 6.2

Two power lines pass through points A and B, as shown in Figure 6.24. The line passing through A bears N 30°E and slopes upward at 20°. The line through B bears N 80°W and slopes upward at 25°. A and B are at the same elevation and 7 m apart.

Determine the minimum distance between the two power lines. The scale is 1:200.

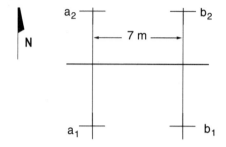

Figure 6.24   Power lines passing through points A and B

What is known?

• A starting point (A and B) on each of the lines
• Bearing and slope of each line.

1. Determine the view necessary to find the required information—the shortest distance between the two lines. The minimum distance will be seen by looking along one of the lines. The view required to measure the minimum distance is one that shows one line as a point view.
2. Show the given information on the plan view. A and B and the bearing are seen. Figure 6.25 shows the plan view.

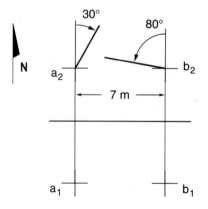

Figure 6.25   Plan view showing bearings

3. Use the slope of line A to construct a view showing the true length. An auxiliary view, 3, is located parallel to line A in the plan view. Line A will be true length in this auxiliary view. The actual length is not important. Figure 6.26 shows the true length of line A.

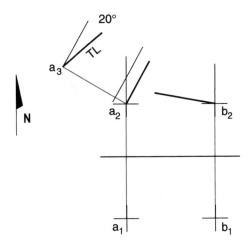

Figure 6.26   True length of line A

4. Line B cannot be projected onto the auxiliary plane until you have established another point on it. Use the slope of B to draw the true length of B. An auxiliary plane, 4, is located parallel to line B in the plan view. The true length of B will appear on this auxiliary. Figure 6.27 shows the true length of line B. Any point, C, can be used.

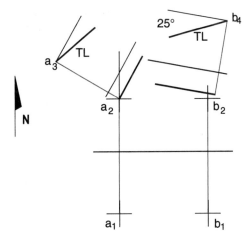

Figure 6.27   True length of line B

5. Project C to the plan view and then project BC to the auxiliary view showing the true length of A. Figure 6.28 shows the two lines on plane 3.

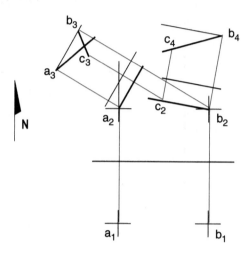

Figure 6.28   The two lines on plane 3

6. Locate an auxiliary view to show the point view of A. The auxiliary plane, 5, is perpendicular to the true length of A. Project A onto auxiliary plane 5.

   The shortest distance between A and B is the perpendicular distance between the point view of A and line B. Figure 6.29 shows this distance.

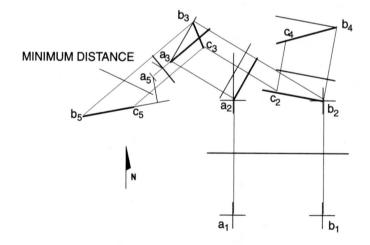

Figure 6.29   Distance between point view of A and line B

Now that you know a little more about the properties of lines, try the following problems.

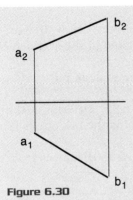

Figure 6.30

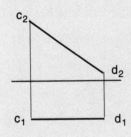

Figure 6.31

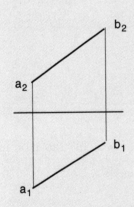

Figure 6.32

Problems

1. Draw a true length of line AB (Figure 6.30) by projecting a) from the plan view b) from the front view and measure the true length. The scale is 1:1.

   A is 14 mm behind the frontal plane and 8 mm below the horizontal plane.

   B is 22 mm behind the frontal plane and 25 mm below the horizontal plane.

2. Draw the point view of line CD (Figure 6.31). The scale is 1:1.

   C is 15 mm behind the frontal plane.

   D is 2 mm behind the frontal plane.

   Both points are 10 mm below the horizontal plane.

3. Draw a point view of line AB, (Figure 6.32).

   A is 10 mm behind the frontal plane and 20 mm below the horizontal plane.

   B is 22 mm behind the frontal plane and 8 mm below the horizontal plane.

4. A line, AB, is 7 m long, bears N 40°E, and slopes down from A at 30°. Draw plan and front views of AB. Choose a suitable scale.

5. A line (CD) starting at any point, C, is 8 m long, bears N 35°W, and slopes downward from C to D at 40°. Draw the point view of CD. The scale is 1:200.

6. A line, BC, bears S 20°W from B and slopes downward at 30°. One end, B, is 700 mm higher than the other. Draw plan and front elevation views and determine the length of the line. The scale is 1:50.

7. Point B is 300 m south and 200 m west of C and is at an elevation of 1200 m (above sea level). It slopes down from B at 25°. Draw the plan and north elevation views of the line and determine the elevation of C. The scale is 1:10000.

8. Find the slope and the length of line AB.

   A is 200 mm behind the frontal plane and 120 mm below the horizontal plane.

   B is 50 mm behind the frontal plane and 200 mm below the horizontal plane.

   The scale is 1:10.

9. Point A must be connected to line BC by the shortest connector. How long is the required connector? Show the connector in the plan and front views.

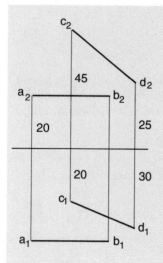

Figure 6.33

B is 150 mm behind the frontal plane and 150 mm below the horizontal. C is 150 mm north, 300 mm east, and at the same elevation as B.

A is 230 mm east and 100 mm below B. The scale is 1:10.

10. Two lines, AB and CD, are shown in Figure 6.33. Determine the length, bearing, and slope of a connector that joins the midpoints of these lines.

C is 15 mm east of A, and D is 40 mm east of A. Distances from the folding line are shown in Figure 6.33.

11. Line BC is 1.5 m long and has an azimuth of 25°. A point view (PV) is 25 mm behind the folding line on the auxiliary view, which shows the line as a point. Draw the missing views of line BC. The layout is shown in Figure 6.34. The scale is 1:50.

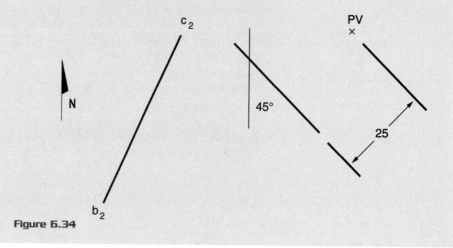

Figure 6.34

Now that you are familiar with the properties of lines, let's move on to look at the properties of planes.

Properties of Planes

You must know how to find the two limiting cases of a plane: the true shape and the edge view. These views, used in combination, can be used to solve any problems involving planes. A minimum of three lines are required to define a plane, but more lines can be added if necessary. The only requirement is that any lines added must be in the plane. The edge view must be found before a true shape can be found.

Edge View of a Plane

A plane will be seen in edge view *if any line* in it is seen as a point. It is often easier to add a line than to use an existing line.

Finding the Edge View of a Plane

Figure 6.35 shows plan and front views of plane ABC. The requirement is to find the edge view of this plane.

The plane can be seen as an edge if any line in it is seen as a point view. As you know, a line must be seen in true length before it can be seen in point view. Any line—AB, BC, or CA—can be seen in true length, then as a point view. The plane can be seen in edge view. This requires two auxiliary views; however, there is a better way.

None of the given lines AB, BC, or CA is true length since none is parallel to the folding line in either view. We will construct a true length line in the plane and then find the point view of this new line. Only one auxiliary view will be required to find an edge view.

1. Draw a line, in the plane, parallel to the folding line in the front view. This line will be true length when projected onto the plan view. Figure 6.36 shows the line added to the front view. The new line is a horizontal, or level, line.

    Start the new line at a given point in the plane (A). Call the other end X. If the line starts at a given point, only one end has to be projected onto the other view.
2. Project point X onto the plan view. It will be on BC. Label the line as true length (TL).

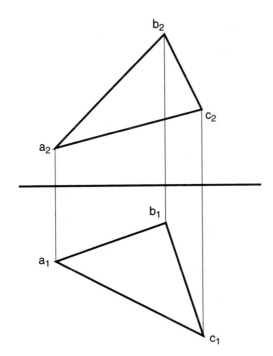

Figure 6.35
Front and plan views of a plane

Figure 6.36
Creating a true length line in a plane

3. Find the point view of AX. Locate a folding line perpendicular to AX and project AX onto the new plane (plane 3). Figure 6.37 shows the point view of AX projected onto plane 3.
4. Project B and C onto plane 3 and join the points to see the edge view of ABC. Figure 6.38 shows the edge view.

A true length line can be created in the front view by drawing a line parallel to the folding line in the plan view. The auxiliary plane showing the edge view can then be taken off the front view.

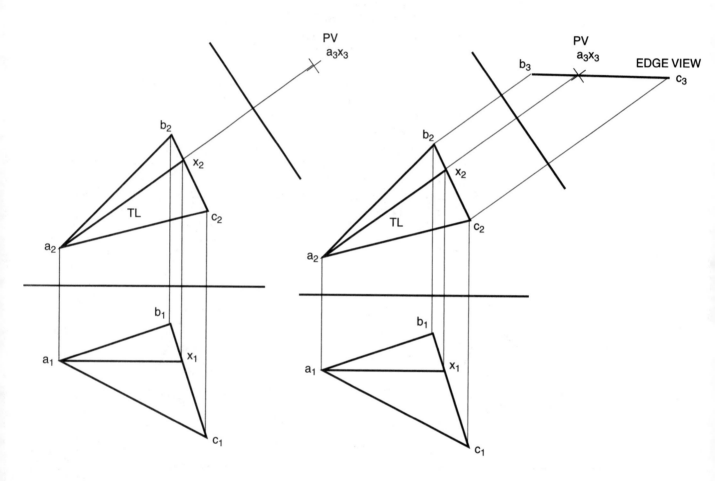

Figure 6.37 Point view of one line in the plane

Figure 6.38 Edge view of a plane

True Shape of a Plane

The true shape will be seen when viewed from a direction perpendicular to the plane. An edge view must be found before the true shape can be drawn. An auxiliary view is then positioned so that the line of sight is perpendicular to the edge view. The true shape will be seen in this auxiliary view.

Finding the True Shape of a Plane

1. Draw the plane in edge view (see Figure 6.39). This is a repeat of Figure 6.38.

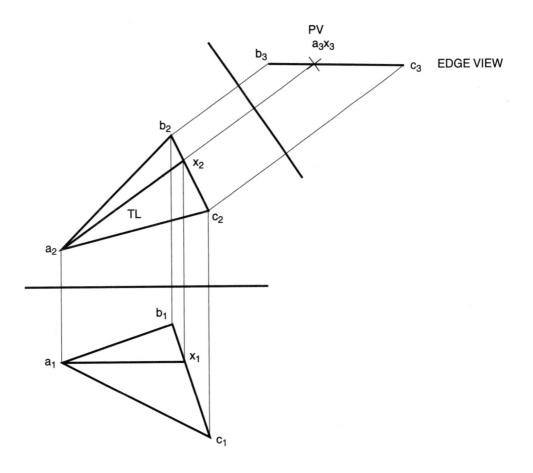

Figure 6.39   Edge view of a plane

2. Place a folding line parallel to the edge view. It can be on either side of the edge view. Figure 6.40 shows the location of the new auxiliary plane.

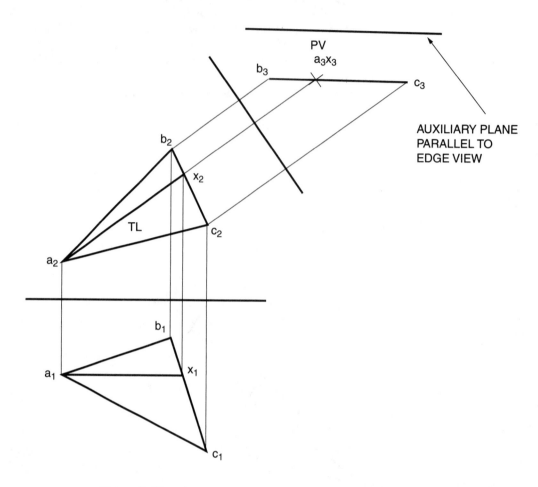

Figure 6.40 Locating auxiliary view to see a true shape

3. Project all points onto auxiliary plane 4 and join them. Figure 6.41
    shows the true shape of the plane.

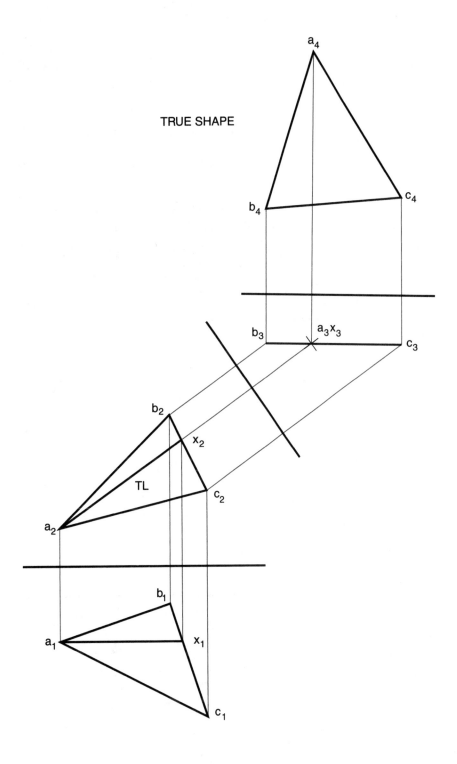

Figure 6.41 True shape of the plane

Slope of a Plane

The slope of a plane is the angle it makes with the horizontal. There can be only one slope. Slope is seen only in an elevation view where the plane is seen in edge view. An elevation view *must* be used because a horizontal plane (with which slope is measured) is seen as an edge only in an elevation view. The true length line used to find the edge view *must* be in the plan view for the edge view to appear in an elevation view.

Finding the Slope of a Plane

Figure 6.42 shows two views of a plane ABC. The requirement is to find the slope of the plane.

1. Create a true length line in the plan view by drawing a horizontal line, AX, in the front view. Project AX onto the plan view where it will be true length. Figure 6.43 shows the true length line in the plan view.
2. Draw the edge view of ABC and measure the angle it makes with the horizontal as shown in Figure 6.44. Any line parallel to the folding line is a horizontal line.

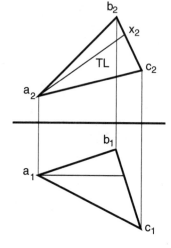

Figure 6.42
Two views of a plane

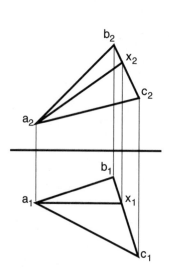

Figure 6.43
True length line created in plan view

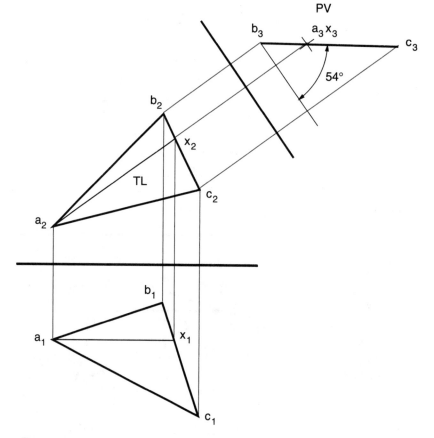

Figure 6.44   Measuring the slope of a plane

The auxiliary view showing the edge view of the plane can be positioned on either side of the plan view.

Now that you know a little more about the properties of planes, try the following problems.

Problems

12. Draw the edge view of plane CDE by projecting from a) the plan view and b) from the front view. A true length line has been drawn in each view. Layout dimensions are given in Figure 6.45. The scale is 1:1.

13. Draw the true shape of plane FGH (Figure 6.46). The edge view, which must be found first, can be projected from either the plan or the front view.

14. Determine the distance between point A and plane BCD. All points are referenced from the lowest point, E. The scale is 1:10.

B is 350 mm west, 100 mm north, and 200 mm above D.
C is 100 mm west, 250 mm north, and 300 mm above D.
A is 250 mm west and 300 mm above D.

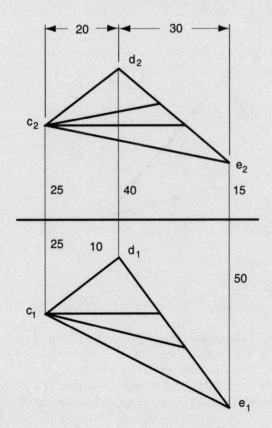

Figure 6.45

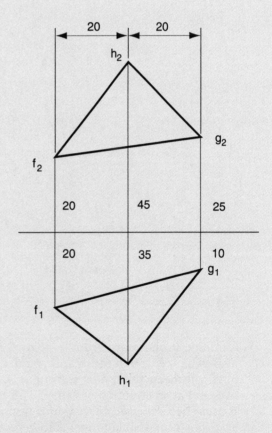

Figure 6.46

15. Determine the slope of plane CDE. All points are referenced from the lowest point. The scale is 1:20.
    D is 700 mm west, 600 mm north, and 450 mm above F.
    E is 1000 mm west, 100 mm south, and 220 mm above F.

16. What is the area of the plane in Problem 14?

17. Line AX is 7 m long, bears N 32°E, and is in plane ABC. Determine the vertical distance between X and point B. The scale is 1:100.
    B is 3 m north, 2.5 m east, and 2 m below A.
    C is 1 m north, 4.5 m east, and 1 m above A.

18. The drawing in Figure 6.47 shows a plane, ABC, and a line, AD. Line AD bears N 70°E and is in the plane ABC.
    B is 100 m east and 150 m north of A.
    C is 300 m east and 70 m south of A.
    D is 300 m east of A.
    The elevations of these points are: A 545 m, B 430 m, and C 670 m.
        a) What is the slope of plane ABC?
        b) What is the elevation of D?
    The scale is 1:5000.

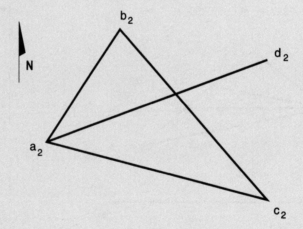

Figure 6.47

19. A mining tunnel starts from A, bears S 65°E, and slopes down at 15°. Another tunnel starts at B, which is 250 m south and 160 m east of A. This tunnel bears N 55°E and slopes down at 20°. It meets the tunnel starting at A at C. A third tunnel starts at D, which is 500 m east of A and at an elevation of 250 m. This tunnel bears S 30°W and slopes upward at 18°. Point B must be connected to this third tunnel by a combination of a tunnel and a vertical shaft. The shortest possible tunnel is to be used.

What is the cost of connecting the tunnels in this way if the cost of drilling tunnels and shafts is $20 000 per metre?

At what elevation does the shaft meet the tunnel from B?

Figure 6.48 shows a layout of the three tunnels. The scale is 1:5000, vertical and horizontal.

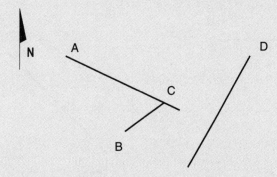

Figure 6.48

20. Plan and elevation views of two power cables are shown in Figure 6.49. The minimum distance between the cables must be 1 m. The only end that can be moved is B, and it can only be lowered. What is the minimum distance that B must be lowered to achieve the minimum clearance? The scale is 1:100.

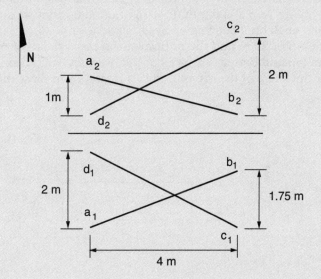

Figure 6.49

21. A plan view showing a 4 m mast near a power line, AB, is illustrated in Figure 6.50. The mast is supported by guy wires, one of which is anchored on the ground at X. The wires are attached to the mast 1 m below the top. The power line slopes down from B at 24°. There must be a minimum clearance of 1 m between the guy wire and the power line. The mast can be moved in a north-south direction to achieve the clearance. How far must the mast be moved to achieve the minimum clearance? The scale is 1:100.

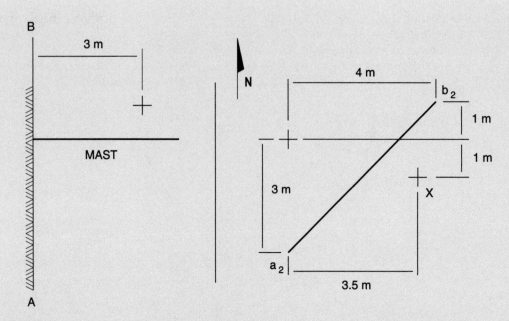

Figure 6.50

22. In the search for minerals and oil deposits, it is usual to drill a hole into the ground and determine where the hole intersects the deposit. By drilling a series of holes, the extent of the deposit can be estimated. A vertical hole drilled into level ground at A (Figure 6.51) hits the top surface of an ore vein at a depth of 40 m and the bottom surface at a depth of 46 m. Another hole drilled at B, on a bearing of N 30°E and a slope of 65°, hits the top surface of the ore vein at a depth of 30 m and the bottom surface at a depth of 60 m (measured along the drill hole). The top and bottom surfaces of the ore vein can be considered parallel.

Determine the thickness of the ore vein in this region from these drilling results. The scale is 1:1000.

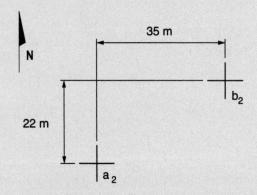

Figure 6.51

23. Two tunnels, AB and CD, are shown in Figure 6.52. They must be connected by a level tunnel bearing N 40°W. What is the length of the level connector? The scale is 1:200.

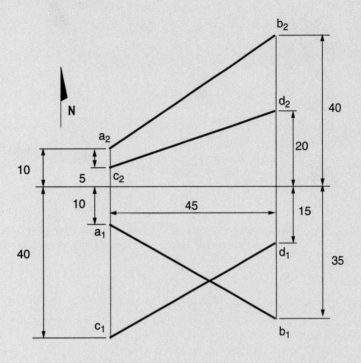

Figure 6.52

Now let's move on to look at another visualization representation, contours.

Contours

A **contour** is a line that indicates a profile on a surface. The most common application of contours is to show lines of constant elevation on the surface of the ground. Figure 6.53 shows contour lines on the surface of a human foot. These contours were determined from photographs and were used as inputs to a numerically controlled machine to make a model of the foot. Each line represents a constant distance above some datum.

Contour lines on a topographical map are lines of constant elevation measured above sea level. If you walked along a contour line, you would neither rise nor fall in elevation. Contour lines on a map show what the ground surface looks like without the use of an elevation view. Figure 6.54 shows a portion of a topographical map. The contour interval and scale must be indicated. In this example, the contour interval is 10 m. Moving from one contour line to the next requires a change in elevation of 10 m.

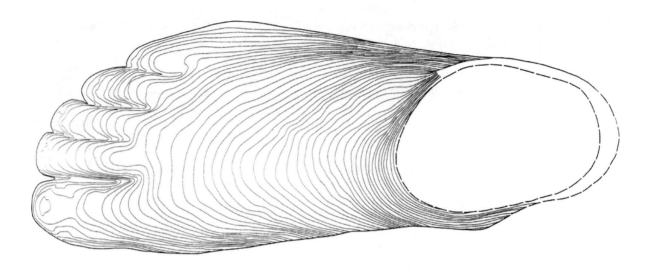

Figure 6.53 Contour lines on the surface of a foot [Courtesy of Prof. James P. Duncan, Retired Department Head, Mechanical Engineering, University of British Columbia]

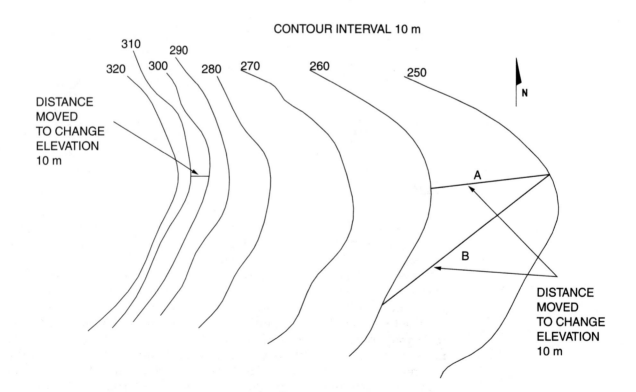

Figure 6.54 Contour lines on a topographical map

If contour lines are close together, as they are on the left side of the map, the slope of the ground is steep. A short run is required to rise or fall 10 m in this region. The wider spacing on the right shows a small slope since the horizontal distance that must be moved to rise or fall 10 m is longer. If route B is followed, the horizontal distance is even longer. If you were in good physical condition, you might take route A. The run for route A is shorter than that for B, indicating that route A is steeper.

Contour lines represent the line of intersection of a horizontal plane with the ground surface. An elevation view showing these planes can be used to show the shape of the ground surface at any location. The process is equivalent to taking a section through the ground. The resulting outline is called a **profile.**

Finding a Profile

Figure 6.55 shows a topographical map.

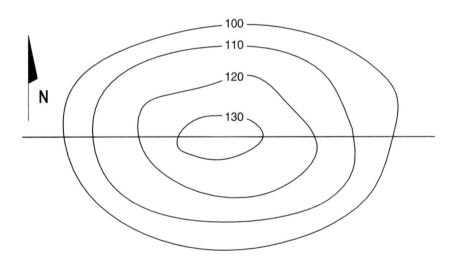

Figure 6.55 Topographical map

1. Draw a line indicating where the profile is taken. The east-west line in Figure 6.55 can be thought of as a vertical plane that cuts through the hill creating a section.
2. Draw an elevation view on which to see the profile. There is no need to draw a folding line. Draw horizontal lines corresponding to the contour interval in the elevation view. Vertical spacing is drawn to scale. Figure 6.56 shows the elevation view and horizontal lines representing the contour intervals.

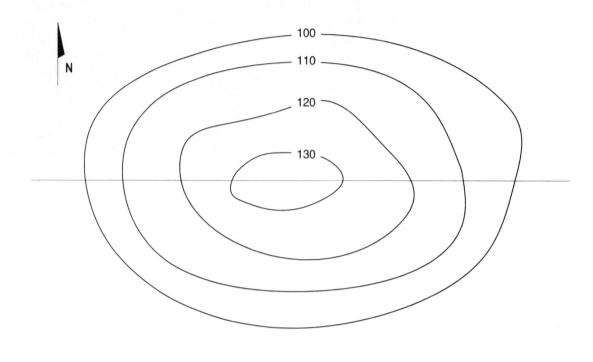

Figure 6.56 Elevation view showing contour intervals

3. Project the points where the east-west line intersects each contour line in the plan view to the corresponding contour in the elevation view (see left-hand side of Figure 6.57). These points are on the ground surface along the east-west line.
4. Join the points with a freehand line representing the ground surface. The symbol for "earth" is also drawn freehand. Be sure the symbol is on the correct side of the line. Figure 6.57 shows the completed profile.

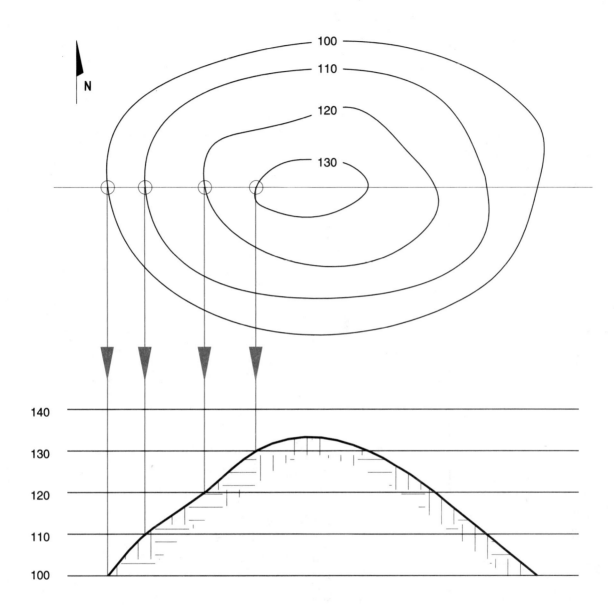

Figure 6.57   Ground profile in an east-west direction

To help you understand the practical application of your knowledge, take a few minutes to look at the following example.

Example 6.3

The topographical map in Figure 6.58 shows a portion of a lake in a mountainous region. A float plane is on the water at the location shown. The airplane must take off on a bearing of N 60°E. The elevation of the lake is 275 m. In the worst case, the plane will lift off the water 20 m from the shore and climb at a rate of 250 m per 1000 m. Determine whether the plane should take off. The contour interval is 5 m. The scale is 1:1000, vertical and horizontal.

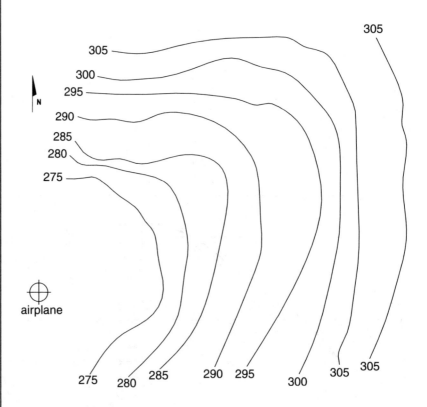

Figure 6.58 Topographical map

Obviously the flight path of the airplane must not intersect the side of the mountain. If it clears the mountain, the takeoff is safe. The path of the airplane must be compared with the profile of the mountain to determine whether they intersect.

What is known?

- Where the airplane leaves the water
- The rate at which the airplane can climb
- The direction in which the airplane takes off.

1. Draw the flight path on the map to show where the profile is being taken. Indicate where the plane leaves the water (20 m from the shore). Figure 6.59 shows this information on the map.

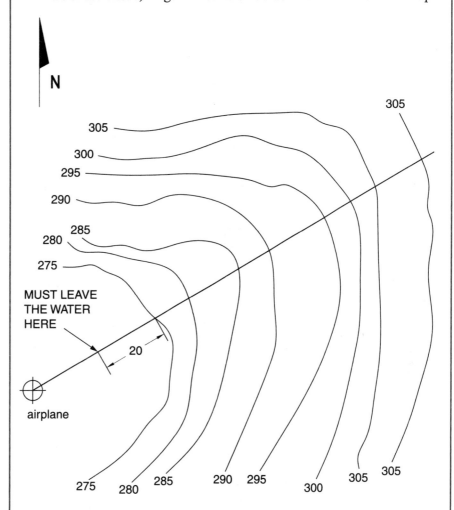

Figure 6.59 Flight path shown

2. Draw an elevation view to show the profile and the flight path. The elevation view is located parallel to the direction in which the airplane flies. Show horizontal lines representing contour intervals on the elevation view (Figure 6.60).

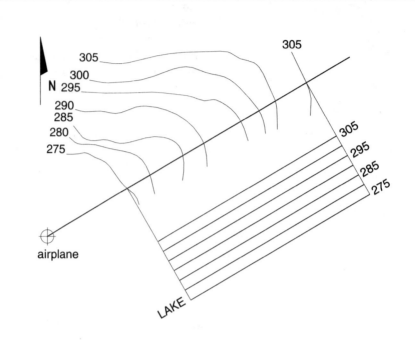

Figure 6.60   Elevation view located to show profile

3. Project the intersection of the flight path and the contour lines in the plan view to the corresponding elevation in the elevation view. Connect the points to show the profile (Figure 6.61)

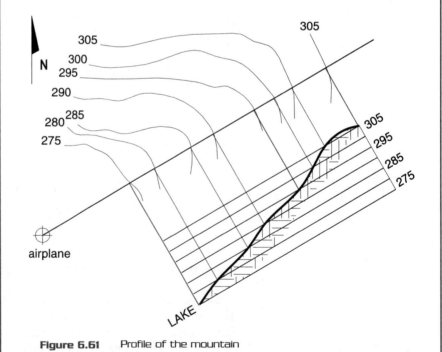

Figure 6.61   Profile of the mountain

4. Draw a line representing the rate of climb on the elevation view. Figure 6.62 shows the flight path compared to the profile of the mountain.

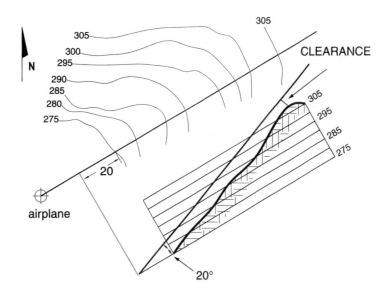

Figure 6.62 Flight path compared with mountain profile

The flight path does not intersect the mountain side, but it is very close (clearance is about 1 m). It would not be safe to take off under these conditions. The profile (drawn freehand) could not be expected to be accurate to within 1 m and elevations do not take trees into account.

Now that you know a little more about contours, try the following problems.

Problems

24. The topographical map in Figure 6.63 shows the proposed location of a dam site at the convergence of two rivers (river A and river B). Water is carried from the reservoir behind the dam to the turbine in the powerhouse by a large pipe called a penstock. The penstock enters the powerhouse on a bearing of due east (point A) and an elevation of 1400 m at the centerline. This section of the penstock is 260 m long and is horizontal. From the west end of this horizontal section, the penstock bears N 20°W and slopes upward at 30°. The penstock has another turn so it meets the dam at B, on a bearing of S 75°E and a slope of 30° (downward from B).

Determine:
- The total length of the penstock.
- The length the penstock that is underground.
- The elevation at which the penstock meets the dam (the elevation of B).

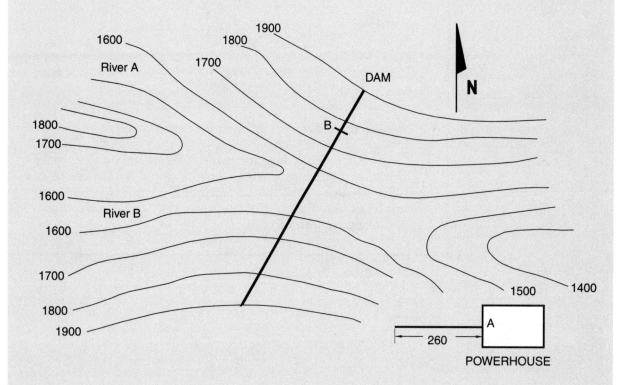

Figure 6.63

25. A ski tow is to go from elevation 470 m to point A at an elevation of 585 m (Figure 6.64). Support towers will be at elevations 470, 510, 530, 550, 570, and 580 m. Towers are vertical and are 10 m high.

    Approximate the cable (at the top of each tower) as a series of straight lines and determine the length of cable required to go from the bottom tower to the top tower.

26. In a region where the ground has a uniform slope, the contour lines are straight parallel lines bearing N 15°W. The horizontal distance between contour lines is 5 m, as shown on the sketch in Figure 6.65.

    A tunnel 30 m long and bearing N 60°E starts at the surface at an elevation of 550 m (point A in Figure 6.65). The tunnel slopes downward from A at 30°.

    A new tunnel must be drilled from the ground surface to the bottom of this tunnel. The new tunnel will slope at 45° on a bearing of N 45°W.

    Determine where the new tunnel must start in order to meet the end of the old tunnel. Specify the drilling location by giving the distances south and east of A. The scale is 1:500, horizontal and vertical.

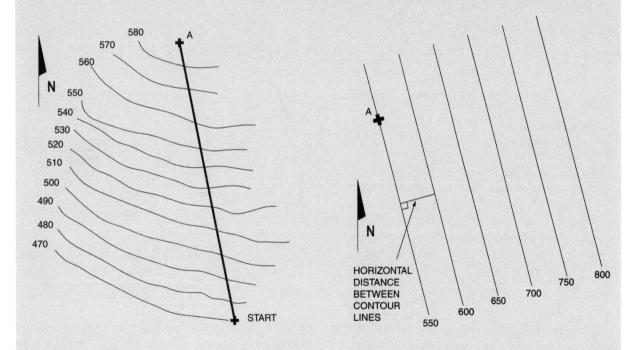

Figure 6.64   **Figure 6.65**

27. A contour map, Figure 6.66, shows a region where a vein of ore meets the surface of the ground. The top and bottom surfaces of the vein are parallel. A vertical hole at A (elevation 512 m) hits the top of the vein at an elevation of 502 m. A vertical hole at B, 34 m east of A, hits the top surface of the vein at 512 m and the bottom at 504 m. A point on the top surface of the vein is at the ground surface at C, 32.5 m north and 26 m east of A. The scale is 1:1000, vertical and horizontal.

Determine the thickness of the ore vein.

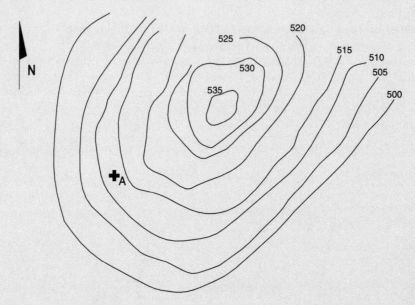

Figure 6.66

28. The contour map in Figure 6.67 shows an area for a proposed ski resort. A ski tow is proposed from A, elevation 130 m, to B, elevation 202 m. The cable must be a minimum of 5 m, perpendicular distance from the ground. Approximate the cable by a series of straight lines and determine the length of cable required to go from A to B. The scale is 1:1000, vertical and horizontal.

A ski run is to start at the location shown and end at a location near A. The slope of the run is 20%. Show a possible run that meets these conditions. The ski run may have sharp turns.

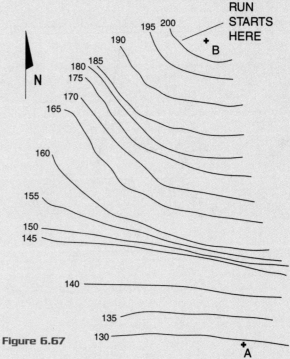

Figure 6.67

29. Figure 6.68 shows a topographical map with hills A, B, and C. Microwave towers are to be situated on the top of hills A and C. If there is no direct line of sight between A and C, another tower must be built on hill B. Using a profile from A to C, determine if a tower is required at B.

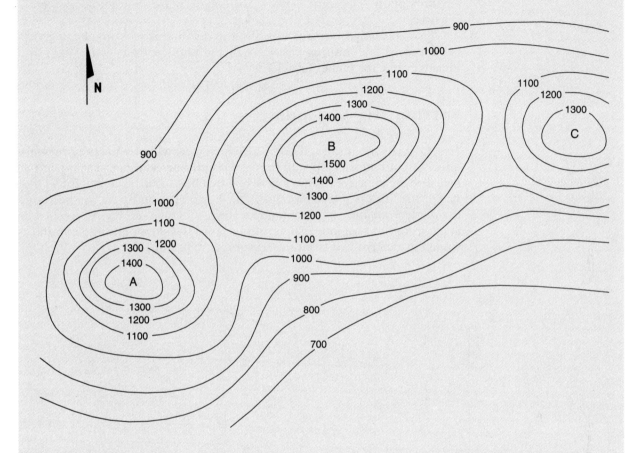

Figure 6.68

Now let's move on to look at some of the 3-D features available on computer programs.

3-D

The 3-D features of a computer program can be used to solve problems similar to those in the first section of this chapter, but instead of drawing auxiliary planes, the viewpoint is changed and the computer screen is the auxiliary plane. The same views—true length, point view, and edge view—are used.

Because computer programs differ in detail, this section takes a generic approach. The limiting cases, true length and point view for lines and the edge view for planes, are described.

Setting the Viewpoint

To set the viewpoint, first determine which view will provide the necessary information and then specify the viewpoint. For example, if you want to measure the shortest distance between two lines, set the viewpoint so that one of the lines is seen as a point. The viewpoint can be set by specifying the angle *from* the x-y plane and *in* the x-y plane (see Figure 6.69). The angle from the x-y plane is determined from the slope. The angle in the x-y plane is determined from the bearing or azimuth.

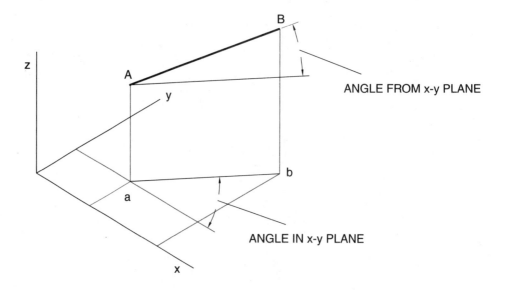

Figure 6.69 Angles required to set the viewpoint

When you enter information on a line or other entity, it is stored as a vector. The program can use this information to determine length, so finding the length is not the problem. You want to set the viewpoint to see the true length on the screen. True length is seen when the line is viewed in a direction perpendicular to it. Figure 6.70 shows two of the many viewpoints that satisfy this condition.

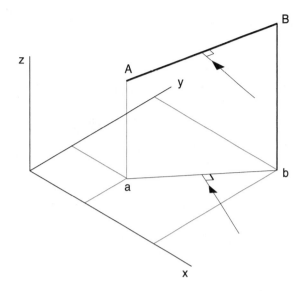

Figure 6.70   Line of sight to show true length

These lines of sight are 90° to the bearing.

Showing the True Length of a Line

Figure 6.71 shows a line with end points A (2, 2, 6) and B (5, 6, 1). You want to specify the angle in the x-y plane so that the line is seen in true length.

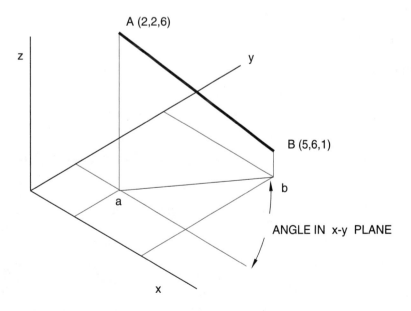

Figure 6.71   Line with end points A and B

You can use your program to give the length and possibly other information. If this is not available, the angle in the x-y plane can be found from the coordinates. The line is 7.07 units long.

1. Determine the angle of the line in the x-y plane. Figure 6.72 shows how this is done using the coordinates. The tangent of the angle in the x-y plane is 4/3 and the angle is 53°. (The bearing is N 37°E.)
2. Determine the angle of the line of sight in the x-y plane. The line of sight is perpendicular to the line. This corresponds to an angle in the x-y plane of -37° or 143°. These angles are shown in Figure 6.73.

    True length can be seen from many other viewpoints, but these are the easiest to set since only the angle in the x-y plane must be determined. The angle from the x-y plane is 0°.

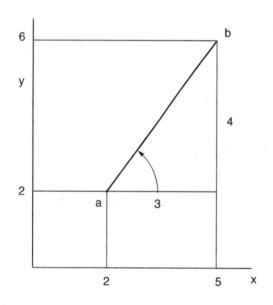

Figure 6.72
Finding the angle of the line in the x-y plane

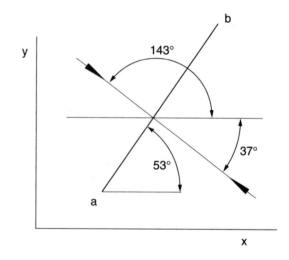

Figure 6.73
Viewpoint angles to see true length

Showing the Slope of a Line

Slope can only be seen when the line of sight is parallel to the x-y plane (the angle from the x-y plane is 0°) and the line is true length. Either of the viewpoints used in Figure 6.74 to show true length will show slope. Figure 6.74 shows the slope of AB.

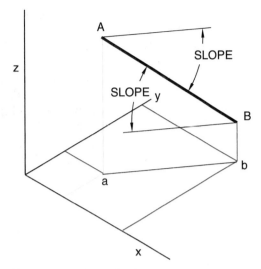

Figure 6.74   Showing the slope of a line

The method of finding slope is the same as that described for finding true length.

Showing a Line as a Point View

A point view can be seen by setting the viewpoint to look along the line. The angle in the x-y plane (direction) and the angle from the x-y plane (slope) must be specified to see a line as a point. Figure 6.75 shows how a point view is seen.

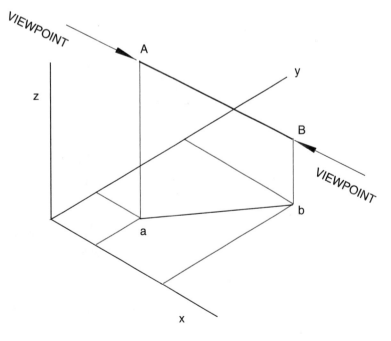

Figure 6.75   Viewpoints to see a point view

Figure 6.76 shows a line with end points A (2, 3, 3) and B (6, 6, 1). You want to specify the angle in the x-y plane and the angle from the x-y plane to see a point view of the line defined by end point A.

1. Find the angle of AB in the x-y plane. Your program may give this information, but you can also find it from the coordinates. The tangent of the angle in the x-y plane is 3/4 and the angle is 37°. Figure 6.77 shows how this is found.
2. Find the slope of AB. The line slopes down from A with a drop of 2 and a run of 5 (the length of AB in the x-y plane). The slope is -22° (down from A). This is shown in Figure 6.78.

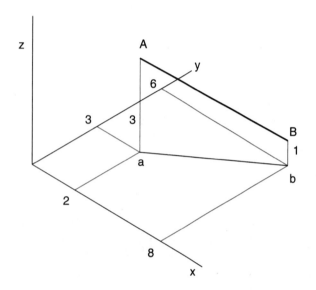

Figure 6.76 Line with end points A and B

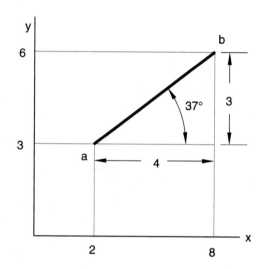

Figure 6.77 Finding the angle of AB

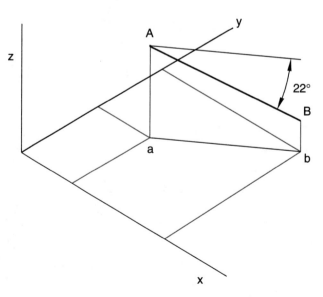

Figure 6.78 Finding the slope of AB

3. A point view can be seen from either end of AB. Figure 6.79 shows both possible angles. The angle in the x-y plane is 37° when viewed from B and 217° when viewed from A.

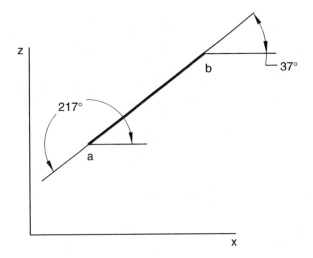

Figure 6.79   View point angles

4. Find the angle from the x-y plane. There are two possible angles depending on the direction in which the line is viewed. When viewed from the low end, B, the angle from the x-y plane is -22° since the viewpoint is below the x-y plane. When viewed from the high end, A, the angle from the x-y plane is 22° since the viewpoint is above the x-y plane. Figure 6.80 shows the angles required to set the viewpoint for each end.

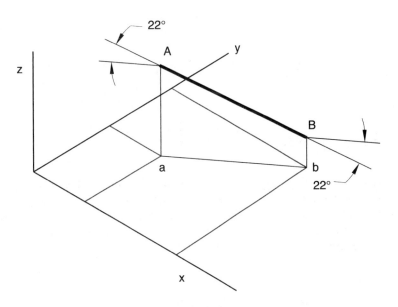

Figure 6.80   Angles to set the viewpoint for each end

Now that you know a little more about 3-D lines and viewpoints, try the following problems.

Problems

The problems in this set involve lines. Part of the problem includes inputting the data.

30. Show the true length of a line AB, with end points A (6, 5, 7) and B (2, 3, 2).

31. Show the slope of the line with end points (1, 3, 3) and (9, 10, 8). Dimension the slope.

32. What is the bearing of a line AB with end points (4, 8, 6) and (10, 4, 3)?

33. Determine the length, slope, and bearing of the line with end points (2, 3, 6) and (8, 7, 1).

34. What is the shortest distance between the lines, AB and CD, defined by:
    A   6, 3, 0?   B   1, 5, 6?
    C   0, 2, 1?   D   7, 8, 8?

    Show and dimension this distance. The shortest distance is the perpendicular distance. Find one line as a point view and measure the perpendicular distance from the point view to the other line.

35. Determine the length of the shortest connection between AB and CD.
    All points are referenced from A. The lines may be extended as required.
        B is 330 mm east, 460 mm north, and 400 mm below A.
        C is 300 mm north, 240 mm west, and the same elevation as A.
        D is 100 mm east, 390 mm north, and 480 mm below A.

Now let's look at using your computer program's features to solve problems relating to planes.

Solving Planes Problems Using a Computer

You would want to view a plane as an edge if you wanted to measure the angle it made with another plane or to determine an intersection point. If any line in the plane is seen as a point view, the plane will be seen as an edge. All that is required is to find a point view of a convenient line in the plane. True shape is seen when a plane is viewed in a direction perpendicular to it. As with lines, both limiting cases are found by setting the viewpoint.

Showing an Edge View and Slope

Slope is seen only when the plane is seen as an edge view and the angle from the x-y plane is 0°. Figure 6.81 shows how slope is seen.

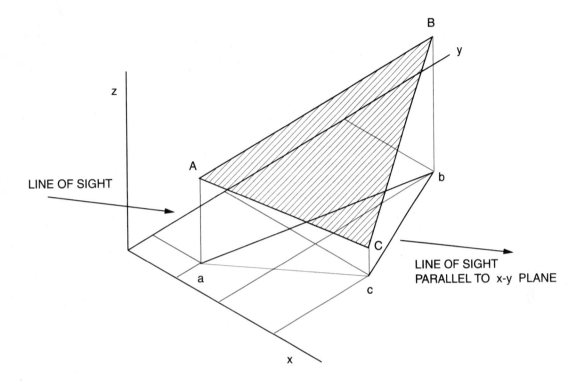

Figure 6.81   Viewpoints to see the slope of a plane

Measure the slope of plane, ABC, defined by:
   A   x = 1   y = 1   z = 2
   B   x = 6   y = 6   z = 6
   C   x = 7   y = 2   z = 1

The plane is shown in Figure 6.82.

The 3-D commands in your computer program must be used to set the viewpoint (or rotate the object). There is no simple way of using coordinates to find the viewpoint as there is with lines. The procedure depends on which program you use and can only be described in general terms.

1. Decide how the plane must be viewed to show slope. Slope is the angle the plane makes with the horizontal plane so slope can be seen only if the line of sight is horizontal. The angle from the x-y plane must be 0° and the plane must be seen as an edge. There are only two viewpoints that satisfy these requirements.
2. Set the angle from the x-y plane to 0° and move the viewpoint left or right until the plane appears as an edge. This is equivalent to rotating the plane about a vertical axis. There are two locations where the edge

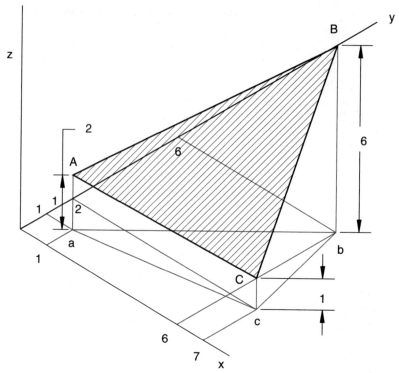

Figure 6.82 Three-sided plane

view appears: 17.25° and 197.25°. Accuracy is limited by how well you position the edge view by hand. Figure 6.83 shows these angles in the x-y plane.

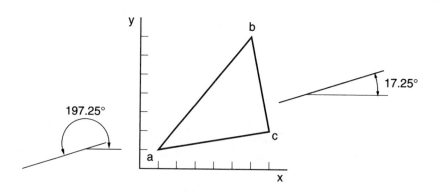

Figure 6.83 Angles in x-y plane to see an edge view

3. Draw a horizontal line so that it passes through the plane and measure the angle the plane makes with this line. Figure 6.84 shows a horizontal line and the slope. How you measure the angle depends on the computer program you are using.

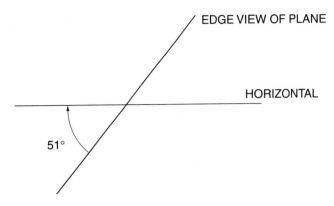

Figure 6.84   Slope of a plane

Showing the True Shape of a Plane

A plane is seen in true shape when the line of sight is perpendicular to it. How this is done depends on the program you are using. AutoCAD, for example, allows you to align coordinates parallel to the plane, giving the equivalent of a plan view. You can then determine area and angles in the plane.

You must specify angles in, and from, the x-y plane to see true shape. The easiest way to find these angles is to first find the slope and the angle in the x-y plane necessary to see slope. The angle from the x-y plane is found from the slope. Figure 6.85 shows how the angle from the x-y plane is related to slope.

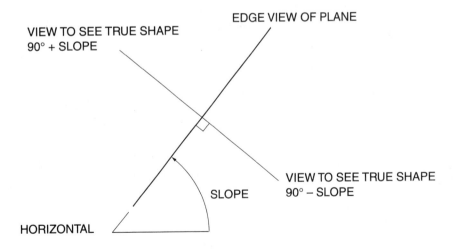

Figure 6.85   Angle from x-y plane to see true shape

The angle from the x-y plane is 90° plus slope or 90° minus slope, depending on which side of the plane is seen.

You can find the angle in the x-y plane by knowing the viewpoint angle, alpha, necessary to see slope. Figure 6.86 shows how these angles are related.

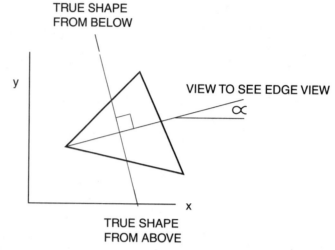

Figure 6.86   Angle in the x-y plane to see true shape

The angle in the x-y plane is either 90° plus alpha or 90° minus alpha, depending on which side of the plane is seen. Viewpoint angles to see the true shape of a plane are:

from above
     in x-y plane   -(90° minus alpha)
     from x-y plane   90° plus slope
from below
     in x-y plane   90° plus alpha
     from the x-y plane   -(90° minus slope)

Figure 6.87 shows the angles when viewed from above.

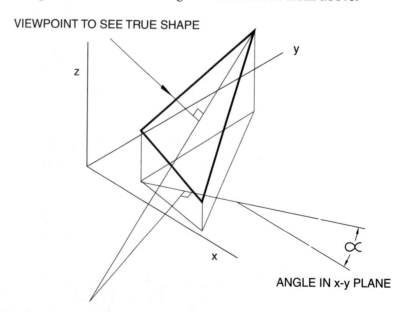

Figure 6.87   Angles when viewed from above

Now that you know a little more about representing planes using your computer, try the following problems.

Problems

36. What is the area of a plane, ABC, defined by A (1, 4, 7), B (7, 7, 4), and C (6, 4, 3)?

37. What is the slope of a plane with orientation defined by A (1, 4, 7), B (7, 7, 4), and C (4, 3, 2)?

38. Determine the shortest distance between a point A (3, 4, 0) and a plane defined by points B (1, 2, 7), C (1, 7, 5), and D (6, 4, 12). The shortest distance will be a perpendicular line from the plane through the point.

39. Determine the distance from the origin to a plane defined by A (3, 3, 10), B (5, 10, 8), and C (10, 5, 4).

40. Dimension the slope of the plane defined by A (4, 1, 8), B (1, 5, 4), and C (8, 9, 1).

41. Determine the area and slope of the plane defined by A (1, 9, 6), B (3, 2, 2), and C (7, 8, 4).

42. Dimension all angles in the plane defined by A (2, 2, 5), B (5, 9, 10), and C (10, 5, 2).

43. A plan view of a plane, ABC, and a line, XY, are shown in Figure 6.88. Locate the point at which the line intersects the plane. Specify the location of the intersection point with reference to the lowest point in the plane. All points are located from the lowest point, C.
    A is 170 mm east, 350 mm south, and 250 mm above C.
    B is 290 mm west, 200 mm south, and 80 mm above C.
    X is 240 mm west, 120 mm south, and 220 mm above C.
    Y is 350 mm south and 80 mm above C.

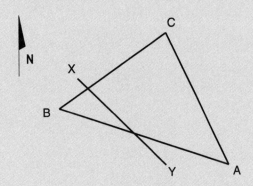

Figure 6.88

44. Referring to Figure 6.89, what angle does the sloping face make with the bottom surface of the block? The object is made from a cube with sides of 750 mm.

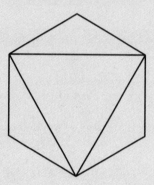

Figure 6.89

45. Figure 6.90 shows an object with several plane surfaces. Determine the angle plane ABC makes with the horizontal.

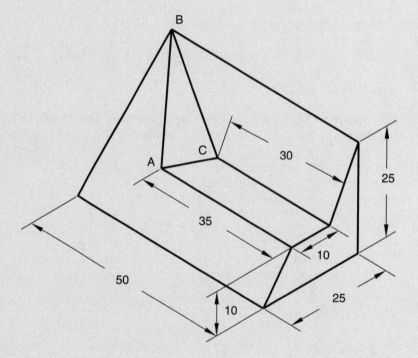

Figure 6.90

46. Determine the area of the plane surface, A-to-H, shown in Figure 6.91. What is the slope of this surface?

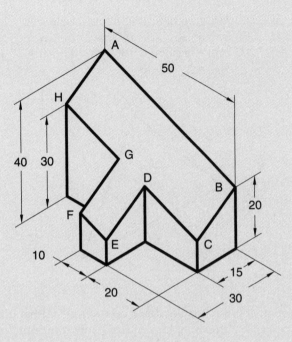

Figure 6.91

47. Figure 6.92 shows a 25 x 25 x 50 mm block with one end cut at an angle. Determine the area and slope of surface ABCD.

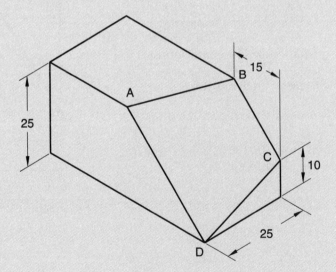

Figure 6.92

48. The line passing through X slopes downward 43° on a bearing of S 52°W and intersects the plane ABC (Figure 6.93). Determine the angle the line makes with the plane.
    The location of all points is given from A.
      B is 3200 mm east, 2600 mm north, and 4800 mm below A.
      C is 5200 mm east, 2600 mm south, and 1600 mm below A.
      X is 6600 mm east, 3200 mm north, and 1800 mm above A.

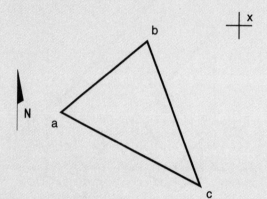

Figure 6.93

There are many cases when there is no database containing information on the problem and the information must entered. Entering the required information then becomes part of the problem.

49. Figure 6.94 shows plan and elevation views of a triangular steel frame, ABC. All points are in the same plane. A power line runs through point X on a bearing of S 45°W and slopes down at 23°. Determine the clearance between the line and the closest frame member. (The wire goes through the center of the frame.)
      B is 6000 m east and 3400 mm lower than A.
      C is 3000 m east and 5200 mm north of A.
      X is 6000 mm east, 4000 mm north, and at the same elevation as A.

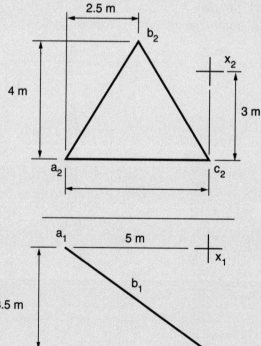

Figure 6.94   3.5 m

50 The centerlines of two pipes, AB and CD, are shown in Figure 6.95. What is the shortest distance between the two pipes?

      B is 3300 mm east, 1400 mm north, 1000 mm higher than A.

      C is 600 mm west, 2400 mm north, 1700 mm lower than A.

      D is 3600 mm east, 400 mm north, and at the same elevation as A.

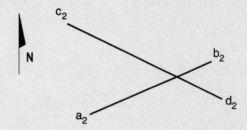

Figure 6.95

51. Figure 6.96 shows an electrical wire near a vertical mast supported by guy wires (not shown). One of the guy wires is anchored to the ground at location X. The other end must be attached to the mast as close to the top as possible. There must be a minimum of 600 mm clearance between the guy wire and the electrical wire. At what height above the ground can the guy wire be attached to the mast under these conditions? The ground is level so the base of the mast and the anchor (at X) are at the same elevation.

      The anchor is 4000 mm east and 2000 mm south of the mast.

      Point B (a point on the wire) is 5000 mm east, 1700 mm north of the mast, and 4100 mm above the ground.

      The electrical wire bears S 35°W from B and slopes down (from B) at 22°.

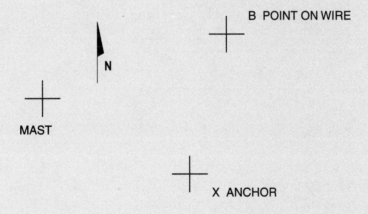

Figure 6.96

52. The orientation of a panel is defined by three points (see Figure 6.97). A pipe passing through X bears N 53°W and slopes up from X at 12°. The end of the pipe must be cut so that it is flush with the wall. What is the angle between the pipe and the panel ?

    B is 2800 mm north, 2000 mm east, and 2200 mm below A.

    C is 700 mm north, 3700 mm east, and 1000 mm below A.

    X is 700 mm south, 3200 mm east, and 2200 mm below A.

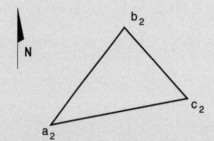

Figure 6.97

53. Figure 6.98 shows the center line of a 200 mm diameter pipe and the edge of a structural support, MN. What is the clearance between the pipe and the support?

    A and B are points on the pipe centerline and M and N are points on the edge of the structural support.

    B is 750 mm south, 300 mm east, and 150 mm above A.

    M is 120 mm south, 230 mm west, and 170 mm above A.

    N is 490 mm south, 150 mm east, and 220 mm above B.

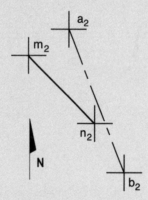

Figure 6.98

54. The drawing in Figure 6.99 shows a connection between two rafters in a roof. (This type of roof is called a hip roof.) The sketch shows where these rafters are located in the roof. The rafter from the corner to the ridge is a hip rafter. The other is called a jack rafter. The end of the jack rafter must be cut so that it meets the hip rafter. Determine the angles on each face of the jack rafter so there are no gaps at the connection. The dimensions of the rafters may be taken as 40 x 90 mm. The vertical face is 90 mm.

The top of the roof is 2000 mm above the top of the wall. The end of the roof is 6 m wide and the slope of the end panel is 1:2.

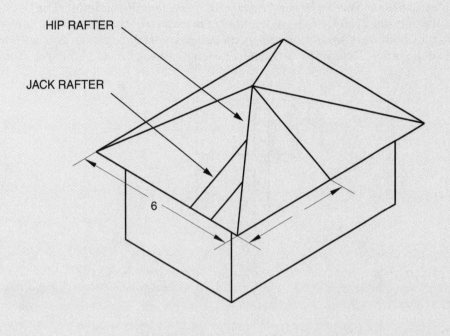

HIP RAFTER

JACK RAFTER

6

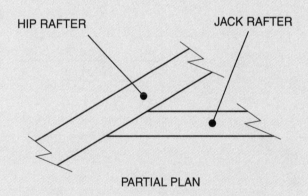

HIP RAFTER   JACK RAFTER

PARTIAL PLAN

Figure 6.99

55. The location, A, where an ore vein hits the ground surface (called an outcrop) is known. The vein is known to be a plane, so finding three points in it will enable the slope to be determined. In order to find the slope, two other points must be found. This is done by drilling two holes and finding the points at which these hit the top of the vein.

    An inclined hole is drilled at M, 50 m west, 40 m south, and 10 m below A. This hole slopes at 45° and has a bearing of S 60°E. The ore vein is struck after drilling 40 m (measured along the hole).

    Another inclined hole is drilled at N, 50 m east, 25 m south, and 8 m below A. This hole bears S 45°W and slopes at 60°. The vein is hit after drilling 60 m (measured along the hole).

    Determine the slope of the vein.

56. One of the components of an earthquake simulator is a vertical column made with an H-section. This must be braced from the floor with a hollow structural section welded to a steel plate on the floor. The brace has a square cross-section 75 × 75 mm. The brace will be welded to the column 2400 mm above the floor. The center of the plate is 900 mm south of the attachment point on the column and 1200 mm east. All dimensions are taken from the centerline of the brace. Determine the centerline length of the brace. Figure 6.100 shows the layout.

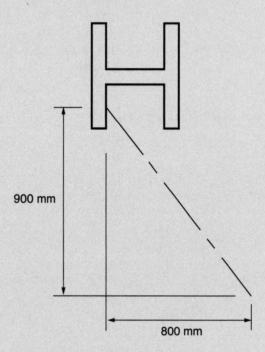

900 mm

800 mm

Figure 6.100

Chapter 7
Intersection and Development

Intersection

Whenever two entities—lines, planes, or solids—meet, some point or line of **intersection** is formed.

When two lines meet, the intersection is a point; two planes meet on a straight line; and the intersection of curved surfaces produces a curved line.

Before pipes can be welded together to make a roof frame, for instance, the ends must be cut so that they can be joined. Before the ends can be cut, the line of intersection must be determined. When flat surfaces intersect, the angle created must be found. Figure 7.1 shows two pipes meeting at 90°. The end of the vertical pipe is cut so it fits over the horizontal one.

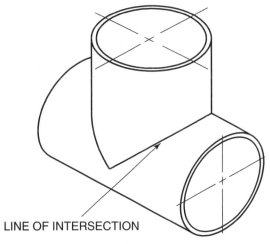

LINE OF INTERSECTION

Figure 7.1   Intersection of two cylinders

A point or line of intersection is found using the same 2-D methods that are used to find the true length and point view of a line. An intersection point can also be found using the 3-D methods described previously. Lines of intersection are formed when two solids are combined using solid modeling techniques; however, it is easier and faster in most cases to use 2-D methods. Computer methods will not be dealt with here.

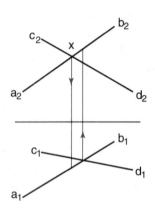

Figure 7.2
Non-intersecting lines

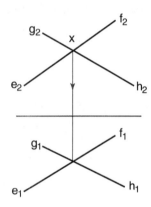

Figure 7.3
Intersecting lines

Intersection of Two Lines

The basis for solving all intersection problems is the application of operations with a line. If you can find the point of intersection of two lines, you can find the line of intersection between two solids. Before you attempt to find a point of intersection, determine whether the lines in question do, in fact, intersect. Two views are needed to ensure that two lines do actually intersect. Figure 7.2 shows two lines, AB and CD, that appear to intersect in the plan view, but, in fact, do not.

If these lines actually intersect, the intersection point, X, will be at the same point on the line in both views. There can be only one intersection between two straight lines. Projecting point X to the front view shows it does not correspond to the apparent intersection point in the front view, so lines AB and CD do not intersect. Lines EF and GH in Figure 7.3 *do* intersect; the intersection point is at the same point on each line in both views.

Intersection of a Line and a Plane

To find where a line intersects a plane, only an edge view of the plane is needed. The intersection can be found using a point view of the line, but more views are required and the intersection point cannot be projected to the line without extra construction.

Example 7.1

Figure 7.4a shows a line that appears to pass through a plane. Find the point at which XY intersects ABC.

1. Draw an edge view of ABC. (Figure 7.4b)
2. Project XY to the auxiliary view. (Figure 7.4c)
3. Project the point of intersection from the auxiliary view to the plan and front views. (Figure 7.4d)

Projection lines are shown only in the view in which they are used.

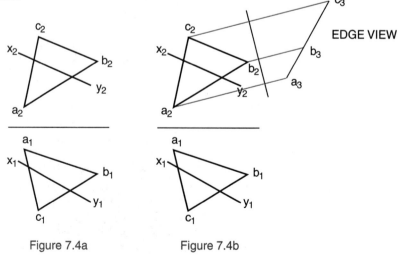

Figure 7.4a   Figure 7.4b

Figure 7.4   Intersection of a line and a plane

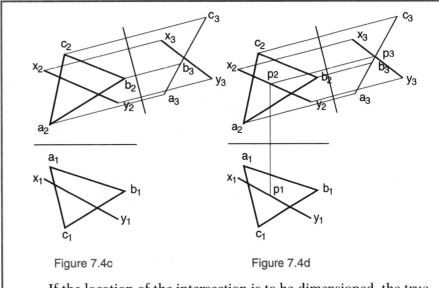

Figure 7.4c   Figure 7.4d

If the location of the intersection is to be dimensioned, the true shape of the plane must be found.

Intersection of Plane Surfaces

The line of intersection formed when two plane surfaces intersect is a straight line.

Example 7.2

Figure 7.5 shows two planes and a line of intersection (BD). The angle between the two planes is called the **dihedral angle**.

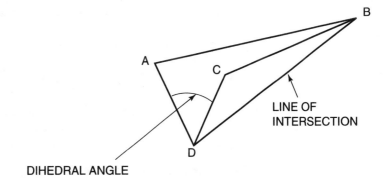

Figure 7.5   Line of intersection and dihedral angle

The line of intersection must be found before the dihedral angle can be measured.

The line of intersection between two planes is in both planes. It can be found by viewing one plane in edge view. Two points on the line of intersection can be seen where this edge passes through the other plane.

1. Draw an auxiliary view showing the edge view of one plane (ABC is used here) and project the other plane (EFG) to this auxiliary. (Figure 7.6)

2. Two points, circled, on the line of intersection are located where the edge view of ABC intersects the sides of EFG. Project these points to the plan and front views. (Figure 7.6b)

Projection lines are shown only in the view in which they are used.

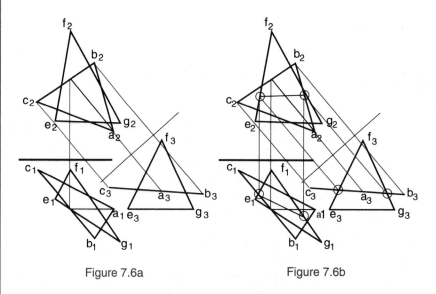

Figure 7.6a   Figure 7.6b

Figure 7.6   Finding the line of intersection between planes

Some parts of each plane will not be visible. However, we are not concerned with visibility; hidden lines are not used for the hidden parts of the planes.

Measuring Dihedral Angle

The angle between two planes (dihedral angle) can be measured if both planes are seen in edge view. We have observed that a plane is seen as an edge if any line in the plane is seen as a point. If the line chosen for the point view is on both planes, as the line of intersection is, both will be seen in edge view, and the dihedral angle can be measured.

Example 7.3

Figure 7.7 shows a roof that must be modified by adding the section containing the plane ABC. A steel support structure must be designed to support the addition, and the angle between panels ABC and ACDE must be known to design the support structure. Determine the angle between the existing roof, represented by ACDE, and the new panel, ABC.

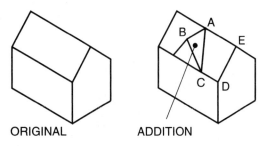

ORIGINAL   ADDITION

Figure 7.7   Addition to a roof

    The dihedral angle can be seen and measured when the line of intersection is seen in point view. The problem becomes one of finding a point view of line AC. A point view of any line can be found using the methods described in Chapter 6.

1. Find the true length of the line of intersection, AC. The true length must be found before a point view can be obtained.

    Figure 7.8 is an auxiliary view showing the true length of the line of intersection. Only one half of the roof is shown.

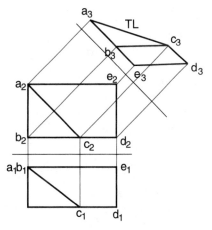

Figure 7.8   True length of the line of intersection

2. Find the point view of the line of intersection. Since a plane appears in edge view, if any line in it is seen as a point, both planes will be seen in edge view in this auxiliary view. Figure 7.9 shows the point view of the line of intersection and the edge views of both planes. The dihedral angle is shown and can be measured.

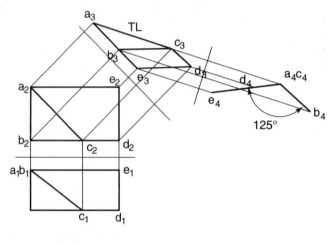

Figure 7.9 Dihedral angle

Now you know a little more about intersections of lines and planes and how to measure dihedral angles, try the following problems.

Problems

1. A line starting at A bears N 37°E and slopes upward at 27°. Another line starts at C, 800 mm west and 900 mm north of A, bears S 73°E, and slopes up at 13°. A and C are at the same elevation. Do these lines intersect?

2. Specify the bearing and slope of a line, starting at D, 8 m east and 5 m south of A, and 8.5 m below C, so that it will intersect plane ABC at the centroid (see Figure 7.10). A **centroid** is like the center of mass of a two-dimensional shape. The centroid of a triangle is at the intersection of the lines joining the vertices to the midpoints of the sides.

   B is 2 m north, 2.5 m east, and 1.5 m below A.
   C is 1.7 m south, 4.5 m east, and 2 m above A.
   D is 2.5 m north, 1 m east, and 1.5 m above A.

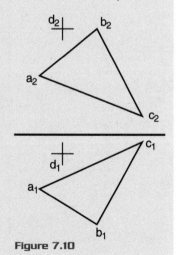

Figure 7.10

3. Refer to Figure 7.11. Line AB bears N 40°W and slopes down from A at 20°. Show the point at which this line intersects plane DEF in all views.

   E is 2 m north, 3 m east, and 1.5 m above D.
   F is 2.3 m south, 2.3 m east, and 800 m below D.
   A is 1 m south, 4 m east, and at the same elevation as E.

4. Refer to Figure 7.12. Determine the point on line XY where it intersects plane ABC.

   B is 1.5 m north, 2.2 m east, and 2 m below A.
   C is 2.8 m north, 2 m west, and 1.7 m below A.
   X is 3.2 m north, 1 m west, and 200 m below A.
   Y is 800 m north, 1 m west, and 1.2 m below A.
   The scale is 1:10.

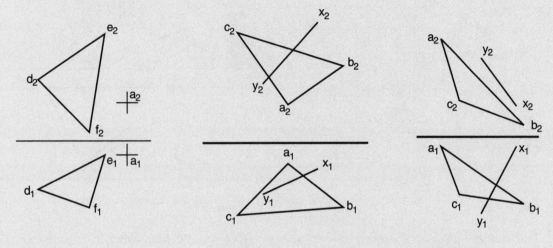

Figure 7.11   **Figure 7.12**   **Figure 7.13**

5. A plan view of a plane, ABC, and an intersection line, XY, are shown in Figure 7.13. Locate the point at which the line intersects the plane. Specify the location of the intersection point with reference to the lowest point in the plane.

   B is 9 m east, 9 m south, and 6 m below A.
   C is 3 m east, 6.4 m south, and 5 m below A.
   X is 8 m east, 7 m south, and at the same elevation as A.
   Y is 4.4 m east, 2 m south, and 7 m below A.
   This defines the orientation of the plane but not the extent.

6. Figure 7.14 shows two intersecting planes. What is the angle between the planes?

   A is 4 m north, 1 m west, and 2 m below C.
   B is 2 m east, 1 m south, and at the same elevation as C.
   D is 2.5 m west, 2 m north, and at the same elevation as C.

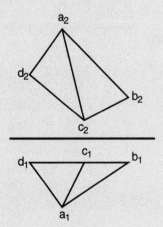

Figure 7.14

7. Details of the planes shown in Figure 7.15 are given below. Determine the bearing and slope of the line of intersection.

   B is 4.7 m east, 500 m south, and 3 m below A.
   C is 2.3 m east, 2.6 m north, and 3.5 m below A.
   D is 4 m east, 2.6 m north, and at the same elevation as A.
   E is 6 m east, 1.6 m north, and 2 m below A.
   F is 1.7 m east, 1 m south, and 4 m below A.

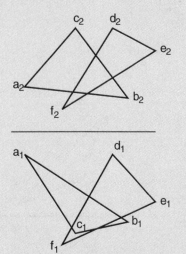

Figure 7.15

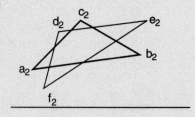

Figure 7.16

8. Determine the dihedral angle between planes ABD and DBC illustrated in Figure 7.16. The scale is 1:200.

   B is 3.4 m north, 3 m east, and at the same elevation as A.
   C is 4 m north, 10 m east, and at the same elevation as A.
   D is 2.4 m south, 8 m east, and 4 m below A.

9. Two planes, ABC and DEF, are shown in Figure 7.17. These planes intersect, but the line of intersection is not shown. Determine the angle between the two planes.

   B is 290 mm east, 40 mm north, and 70 mm above A.
   C is 130 mm east, 130 mm north, and 100 mm below A.
   D is 70 mm east, 100 mm north, and 100 mm above A.
   E is 310 mm east, 130 mm north, and 100 mm below A.
   F is 30 mm east, 50 mm south, and 70 mm below A.

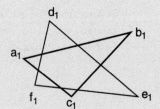

Figure 7.17

10. Figure 7.18 shows two intersecting planes. What is the angle between the planes? Points B, C, and D are measured from A.

    B is 250 mm east, 100 mm south, 150 mm above A.
    C is 50 mm east, 100 mm south, and 100 mm above A.
    D is 50 mm west, 70 mm south, and 200 mm above A.

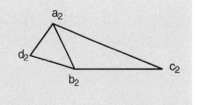

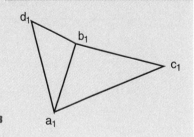

Figure 7.18

Now let's move on to look at the intersection of a plane and a solid.

Intersection of a Plane and a Solid

Conic sections are examples of the intersection of a plane and a solid. The line formed on the surface where the plane cuts through the cone is the line of intersection. It is important to remember that this line is on the surface of the cone. The enclosed shape (which may not be closed, as in the case of a parabola) depends on how the plane passes through the cone. A horizontal plane, parallel to the base, results in a circle; a vertical plane through the vertex results in a triangle; a vertical plane that does not pass through the vertex results in a hyperbola. These shapes are illustrated in Figure 7.19

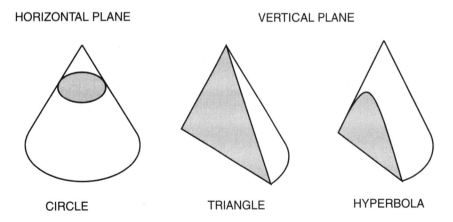

HORIZONTAL PLANE

VERTICAL PLANE

CIRCLE

TRIANGLE

HYPERBOLA

Figure 7.19 Conic sections

If a solid is intersected by a plane, some shape will be formed and there will be a line of intersection on the surface of the solid. This line is in the plane and on the surface. The shape depends on how the plane passes through the solid.

The line of intersection formed when a plane intersects a rectangular prism is a straight line on each face of the prism. Figure 7.20 shows a plane intersecting a rectangular prism parallel to the axis. The line of intersection forms a rectangle. An irregular shape with straight sides results when a plane passes through a rectangular prism at an arbitrary angle.

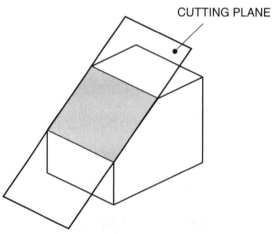

CUTTING PLANE

Figure 7.20   Intersection of a plane and a rectangular solid

When a plane intersects a plane surface, it is necessary to find only two points on the line of intersection with each surface. These points are usually the end points on each side. When a plane passes through a curved surface, more points must be found. Figure 7.21 shows a pipe passing through a wall (an example of a cylinder intersecting a plane).

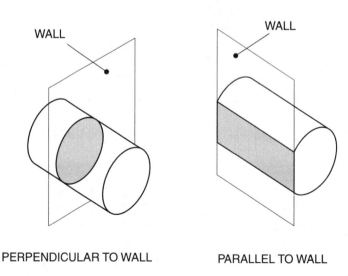

WALL   WALL

PERPENDICULAR TO WALL   PARALLEL TO WALL

Figure 7.21   Intersection of a plane and a cylinder

The shape enclosed by the line of intersection depends on the angle at which the pipe meets the wall. If the pipe meets the wall at a right angle, it will be a circle; if it meets at an angle, it will be an ellipse; if the plane is parallel to the axis of the cylinder, a rectangle is formed.

Finding a Line of Intersection

The following are practical examples of when a line of intersection must be found:

- If a hole is to be made in a wall, the size and shape must be known. A circle is an easier shape to cut on a wall than an ellipse (particularly if the wall is concrete), but it is not always possible to meet the wall at 90°.
- If supports are required, the angle between two panels on a roof must be found so that the supports can be designed.
- If two pipes are joined by welding, the end of one or both (depending on how they are joined) must be cut so that they meet correctly (assuming they are not joined end-to-end).

If a line on one surface intersects a line on another surface, the lines must intersect on the line of intersection between the two surfaces. A line of intersection between two solids can be determined by finding the intersection of a line on one surface with a corresponding line on the other surface. The problem becomes one of finding the intersection of two lines.

Lines can be created on the surface of a solid by passing cutting planes through it. Figure 7.22 shows a cylinder cut by a plane parallel to the axis. The plane meets the surface in two places and creates two lines on the surface; one on the top and one on the underside. These lines appear as point views when viewed from the end of the cylinder. Any number of lines can be created on the surface by passing more cutting planes through it. The reason for using cutting planes to create lines on the surface will become apparent when dealing with the intersection of curved surfaces.

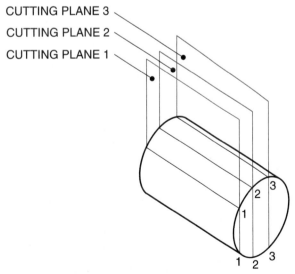

Figure 7.22 Creating lines on a surface using cutting planes

Intersection of a Cylinder and a Plane

Example 7.4

Figure 7.23 shows a large duct passing through a wall. The line of intersection must be found so the hole can be cut. The hole will be an ellipse, since the duct meets the wall at an angle.

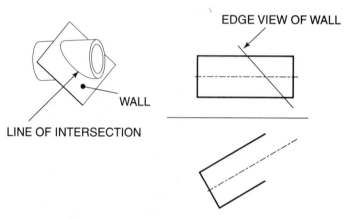

Figure 7.23   Intersection of a pipe and a wall

1. Create lines on the cylinder surface using cutting planes parallel to the cylinder axis. These planes create lines on the top and bottom surfaces, and these lines intersect the wall. In this instance, it does not matter where the cutting planes are. Figure 7.24 shows the duct with three cutting planes, CP1, CP2, and CP3, parallel to the axis. The points at which these lines intersect the wall are identified as 1, 2, 3. These points will be projected to the front view.

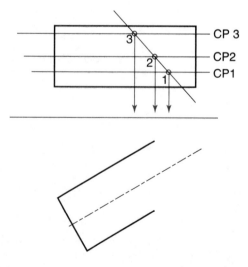

Figure 7.24   Intersection of surface lines with a wall

2. Draw an auxiliary view looking along the axis of the duct. This will enable you to locate the lines around the circumference. Figure 7.25 shows a point view of the duct axis and the duct (which appears as a circle). The auxiliary view is drawn following the procedure for finding the point view of a line outlined in Chapter 6.

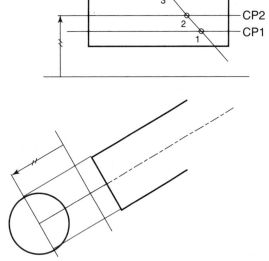

Figure 7.25   View looking along the duct axis

3. Locate the cutting planes in the auxiliary view. They are located using distances from the folding line taken from the top view. The surface lines are seen as point views in the auxiliary view. (see Figure 7.26). The numbers correspond to the cutting planes used to create the lines.

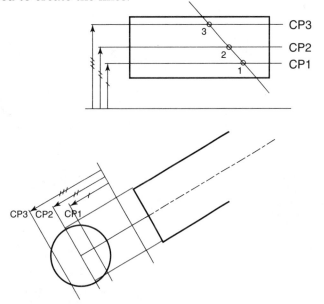

Figure 7.26   Locating surface lines in an auxiliary view

4. Project surface lines from the auxiliary view to the front view. Figure 7.27 shows these lines in the front view.

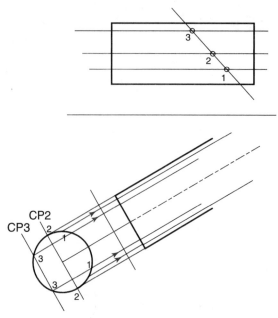

Figure 7.27   Surface lines located in the front view

5. Project the points at which the surface lines intersect the wall to the corresponding line in the front view. Figure 7.28 shows these points on the line of intersection. It is obvious that more points are required to define the line of intersection.

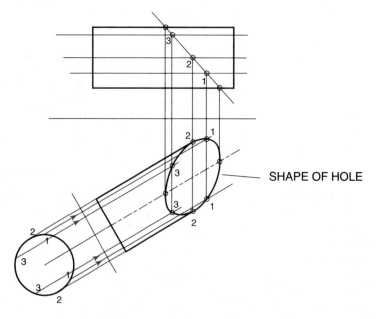

Figure 7.28   Intersection points projected to the front view

6. Create more surface lines around the circumference of the duct and find the intersection points in the front view using the same procedure. Any number of surface lines can be created by using more planes. Figure 7.29 shows the shape of the hole in the wall.

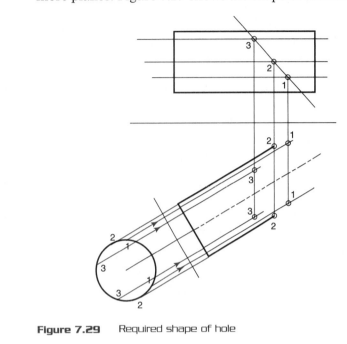

Figure 7.29   Required shape of hole

Now that you know a little more about intersections of planes and solids, and cylinders and solids, try the following problems.

Problems

11. Find the area of the rectangle formed when a 100 mm diameter cylinder is cut by a plane, parallel to the cylinder axis and half way from the axis to the outside (at 1/2 the radius). Find the area per unit length.

12. The grain bin shown in Figure 7.30 must be modified to allow all grain to flow out of the bin. Grain remaining in the bin attracts mice. By experiment it was determined that if the line of intersection between the end panel and the side panel(s) is 35°, grain will not "hang up" in the corner. (The line of intersection slopes at 35°, not the panel.) Determine the dihedral angle between the end panel and the side panel. The angle is the same on both sides.

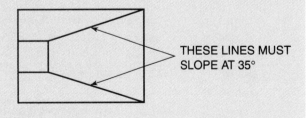

THESE LINES MUST SLOPE AT 35°

Figure 7.30

The top of the bin is 1800 mm wide and 2500 mm long. The hole in the bottom is centered in the bin and is 600 x 600 mm. The top is 1400 mm above the bottom.

13. The roof of the building shown in Figure 7.31 is made with six identical panels. The angle between the panels (dihedral angle) must be known in order to design the roof. The angle is the same for all panels.

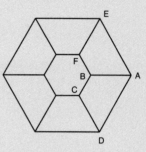

Figure 7.31

   Point A is used as a reference. Since the shape is a hexagon, the included angle between sides is 120°. Size is given knowing the diameter of an enclosing circle.

   The hexagonal roof is contained in a circle with a diameter of 14 m.

   Point B is 5 m west and 3200 mm higher than A.
   Points D and E are at the same elevation as A.
   Points C and F are at the same elevation as B.

14. A vent line must go through a floor and then through a vertical wall. The hole in the floor is to be 400 mm diameter, and the line will pass through it at an angle. A hole must be cut in the wall to accommodate the cylinder. An oblique cylinder will be used because a right cylinder will not pass through a circular hole at an angle. An oblique cylinder has a circular cross section parallel to the base but an elliptical shape when viewed along the axis. (An oblique cylinder is shown in Figure 1.8.) Plan and elevation views are shown in Figure 7.32. Draw the

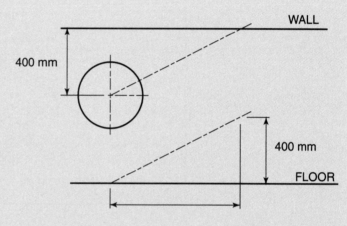

Figure 7.32

shape of the hole required in the wall. The hole will be seen in true shape in the front view. (Find the point at which a line on the surface of the cylinder intersects the wall. Project the intersection point to the corresponding line in the front view.)

15. Find the true shape of the ellipse formed when a cone is cut by a plane passing through it at 30° to the horizontal. The diameter of the base of the cone is 100 mm and the height is 100 mm. The plane passes through the axis of the cone half way between the base and the vertex.

16. A plane passes through a cone such that it intersects the midpoint on the axis and the base at a distance of 25 mm from the axis (1/2 the radius). Draw the true shape of the parabola formed. The cone is 100 mm high and the base diameter is 100 mm.

Now let's move on to look at the intersection of curved surfaces.

Intersection of Curved Surfaces

Lines created on both surfaces for the purpose of finding a line of intersection cannot be drawn randomly. You must be sure that they actually intersect. If two lines are in the same plane, they must intersect unless they are parallel. As you may recall, you use cutting planes to create surface lines to ensure that a line on one surface corresponds to a line on the other surface.

Figure 7.33 shows two intersecting cylinders. A cutting plane is used to create surface lines on both cylinders. Since the lines on the surface of both cylinders are in the same plane, we can be sure they intersect.

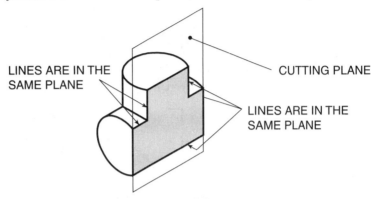

Figure 7.33 Lines on two cylinders created by one plane

Example 7.5

Figure 7.34 shows two pipes that are to be welded together at 90°. The end of the vertical pipe (A) must be cut so it can be joined to the horizontal pipe (B). Determine the line of intersection on the vertical pipe.

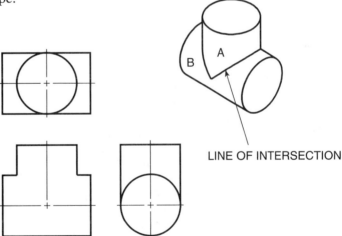

Figure 7.34 Pipes meeting at 90°

1. Create intersecting lines on the surface of each pipe by passing vertical planes through both pipes. The cutting planes used to create surface lines are seen as edge views in the plan view. Figure 7.35 shows the cutting planes in the plan view. Only one cutting plane is shown on the isometric.

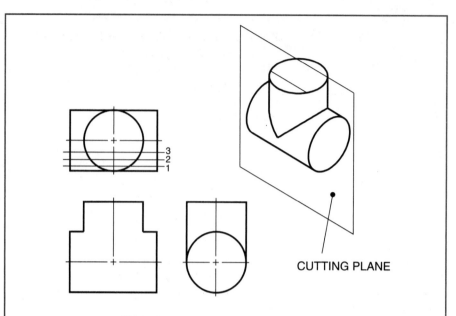

Figure 7.35 Vertical cutting planes added

2. Project vertical lines on the surface of pipe A from the plan view to the front view (Figure 7.36). Now find corresponding lines on pipe B.

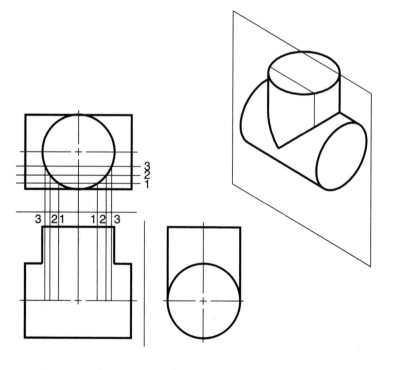

Figure 7.36 Surface lines on vertical cylinder

3. Use the distance from the folding line in the plan view to locate cutting planes in the right-side view. Surface lines on pipe B are seen as point views in this view (see Figure 7.37). Surface lines are only required on the top of horizontal pipe.

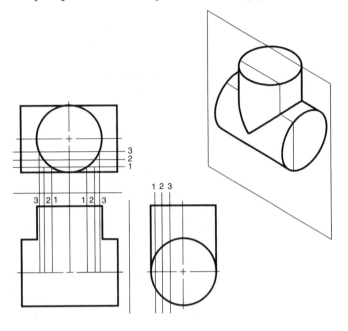

Figure 7.37   Locating surface lines on pipe B

4. Project the surface lines from the right-side view to the front view (see Figure 7.38).

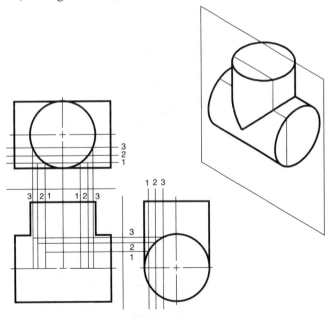

Figure 7.38   Locating surface lines in the front view

5. Identify the intersection points of corresponding lines in the front view. Figure 7.39 shows the intersection points.

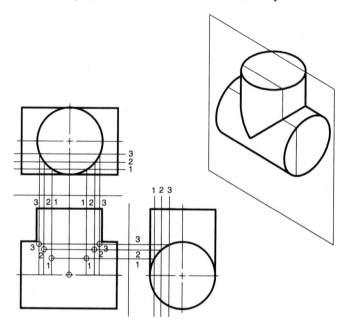

Figure 7.39 Intersection points of surface lines

6. Join the intersection points with a smooth curve. Figure 7.40 shows the line of intersection.

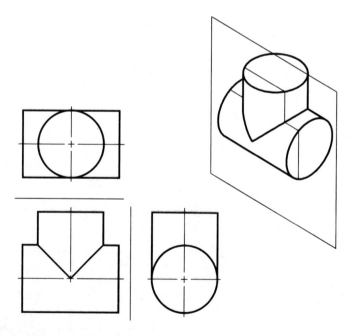

Figure 7.40 Completing the line of intersection

Intersection of a Cylinder and a Cone

Example 7.6

A solid can be separated from a liquid with a simple device called a cyclone separator. This is an inverted cone with an inlet line on the side and outlet lines at top and bottom. Fluid (gas or liquid) and solid enter the side of the separator in a tangential direction. Solid particles are thrown to the side and fall to the bottom, and the fluid flows out through an outlet on the top of the cone. Separation efficiencies of over 90 percent can be attained. Figure 7.41 shows the arrangement of inlet and outlet lines. Find the line of intersection so that a hole can be cut in the body and the inlet line connected. Draw the line of intersection between the inlet line and the body.

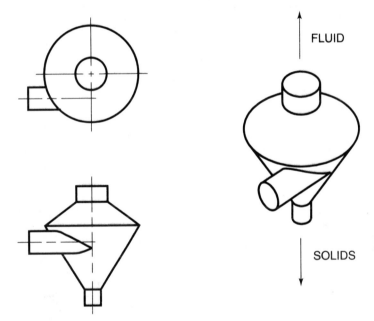

Figure 7.41 Cyclone separator

1. Use horizontal cutting planes passing through the inlet line and the body to create intersecting lines on each surface. Figure 7.42 shows the edge view of the cutting planes in the front view. The top portion of the separator has been removed for clarity.

   Horizontal cutting planes are used instead of vertical ones because the resulting shapes on the conical body (circles) and the round inlet line (straight lines) are easy to draw. Vertical cutting planes would result in straight lines on the inlet line and parabolas on the body. Circles are easier to draw than parabolas.

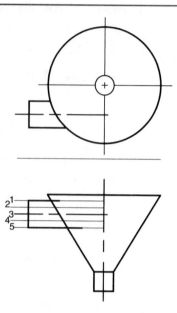

Figure 7.42   Horizontal cutting planes used to create surface lines

2. In the plan view, draw circles corresponding to each cutting plane. The radius of each circle is obtained from the front view. Figure 7.43 shows the circles on the body corresponding to each cutting plane. You are looking down on the inside of the conical body.

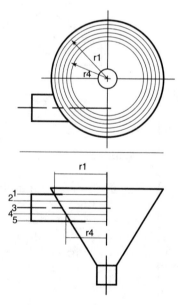

Figure 7.43   Surface lines on the conical body

3. Draw an auxiliary view looking along the axis of the inlet line. The center of the inlet line is located using the distance from the folding line seen in the front view. This view enables surface lines to be located on the inlet line. Figure 7.44 shows this step.

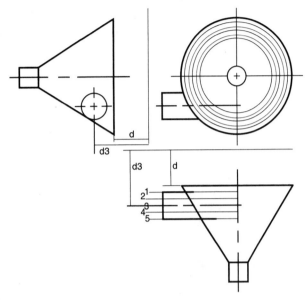

Figure 7.44   Auxiliary view showing the inlet line

4. Locate the cutting planes in the auxiliary view using the distance from the folding line seen in the front view. Lines on the cylinder are seen in point view on the auxiliary view. Figure 7.45 shows these point views on the inlet line. Cutting planes 1 and 5 are at the top and bottom of the cylinder and are not shown in the auxiliary view.

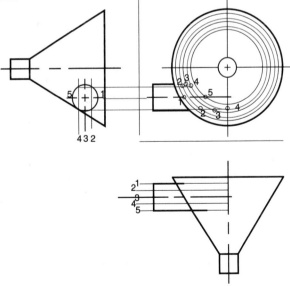

Figure 7.45   Surface lines on the inlet and body

5. Project the surface lines on the inlet from the auxiliary view to the plan view. They will intersect corresponding circles on the body and define the line of intersection in the plan view.

6. Project intersection points from the plan view to corresponding planes in the front view. These points define the line of intersection in the front view. (Figure 7.46)

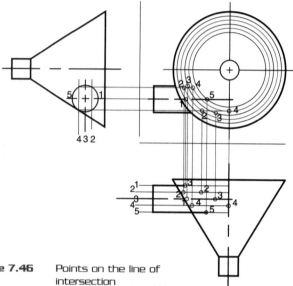

Figure 7.46   Points on the line of intersection

7. Join the intersection points to get the line of intersection in the plan and front views as shown in Figure 7.47.

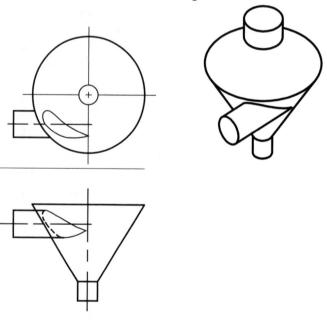

Figure 7.47   Line of intersection

Intersection of a Plane and a Sphere

Example 7.7

Figure 7.48 shows the plan view of a sphere cut by a vertical plane, XX. Draw the line of intersection in the front view.

Any plane passing through a sphere will create a closed curve on the surface. The curve could be a circle or an ellipse, depending on the location of the plane. In this example, the line of intersection will be an ellipse.

You will find the line of intersection in the same manner as in the previous problems, by determining the intersection of lines on the surface of the sphere with the plane.

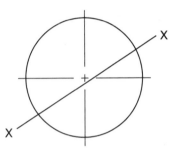

Figure 7.48
Plan view of sphere cut by
vertical plane

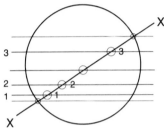

Figure 7.49
Vertical planes used to
create surface lines

1. Pass vertical cutting planes through the sphere to create lines on the surface. These surface lines will appear as circles in the front view. Edge views of these cutting planes are seen in Figure 7.49. The intersection of the surface lines with the plane are numbered (only three of the planes are numbered).

   You could use horizontal planes, but you would first have to draw the front view of the sphere, whereas you can place vertical planes in the plan view, so less work is involved.

2. Draw circles corresponding to the cutting planes in the front view. The radius of each circle is found from the plan view, as shown in Figure 7.50.

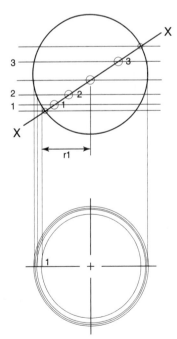

Figure 7.50   Circles corresponding to the cutting planes

3. Project the intersection points from the plan view to the corresponding circle in the front view. There are intersection points on the top and bottom halves of the sphere. Figure 7.51 shows the points defining the line of intersection.

4. Join the intersection points to give the line of intersection. Figure 7.52 shows the line of intersection.

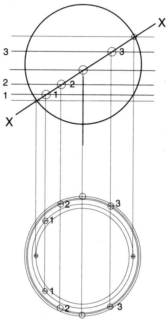

Figure 7.51   Projection of intersection points from plan view to front view

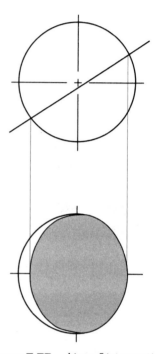

Figure 7.52   Line of intersection

The procedure for finding a line of intersection is the same for all cases: Find a point, or points, where a line on one surface intersects a corresponding line on the other surface. Lines on each surface are created by passing a plane through each surface. This ensures that the lines actually intersect. A series of intersecting lines gives points on the line of intersection.

Flat Surfaces

Rectangular ducts are common in heating, ventilating, and air conditioning applications. Many structural steel shapes are rectangular. If a rectangular section (or other shape made up of plane surfaces) meets a curved surface, the technique for finding the line of intersection is the same as that for curved surfaces, but is usually simpler.

Example 7.8

Figure 7.53 shows a round column supported by a rectangular support. Determine the line of intersection between the top and bottom surfaces of the support and the column.

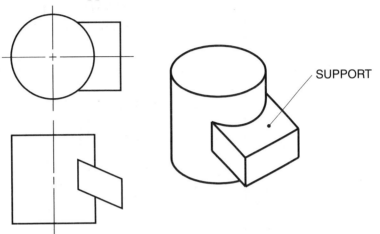

Figure 7.53   Cylindrical section with a rectangular support

The line of intersection can be seen in the plan and front views, but it is not a true shape in either. We will draw the true shape of the line of intersection so it can be cut.

1. Choose points on the line of intersection in the plan view. Intersection points are circled in Figure 7.54.

2. Project these points on the top (and bottom) surfaces in the front view. Only the top surface is shown in Figure 7.55

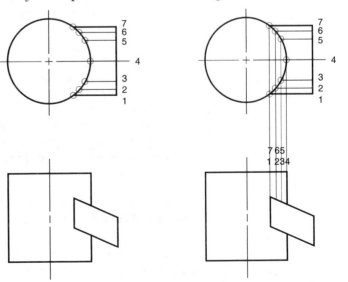

Figure 7.54
Points chosen on the line of
intersection

Figure 7.55
Points on line of intersection found
in front view

3. Draw the true shape of the top surface of the support. The true shape of the bottom surface can also be seen in this view, but is not shown. Since both top and bottom surfaces are seen as edge views in the front view, the true shape can be projected from them. Lines locating intersection points are located from the side of the support in Figure 7.56.

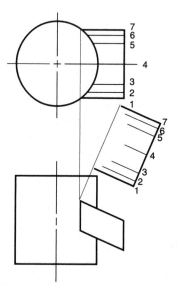

Figure 7.56
True shape of top surface

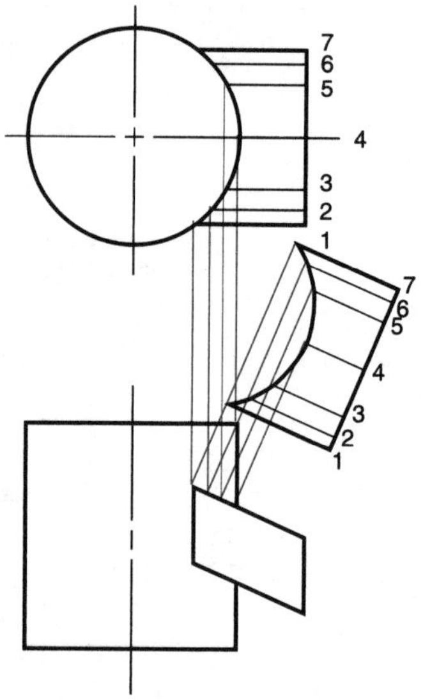

Figure 7.57
True shape of the line of intersection

4. Project intersection points to the true shape view of the top surface. Points can be located using the distance from the front of the support. The line of intersection on the bottom surface is not shown in Figure 7.57.

If the top and bottom surfaces are cut on the line of intersection, they will fit tightly against the column and can be welded.

Now that you know a little more about the intersection of curved surfaces, try the following problems.

Problems

17. Figure 7.58 shows a 50 mm diameter cylinder intersecting a 100 mm diameter cylinder at 90°. Draw the line of intersection between these cylinders.

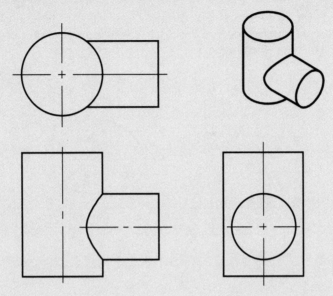

Figure 7.58

18. Figure 7.59 shows a portion of a support for a pier. The frame is to be made from 300 mm diameter tube. One member slopes at 30°. Draw the line of intersection between the tubes.

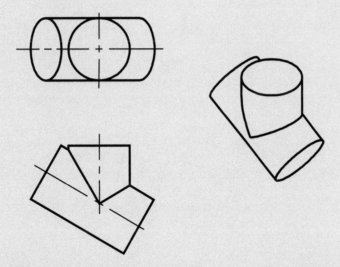

Figure 7.59

19. A horizontal cylinder (50 mm diameter) meets a cone (height 100 mm, base diameter, 100 mm) such that the sides of the cylinder are tangent to the cone. Draw the line of intersection.

20. Figure 7.60 shows plan and front views of the end of a heat exchanger. A 300 mm diameter discharge line leaves the end vertically in the location shown. The end of the heat exchanger is a 700 mm diameter hemisphere. Draw the line of intersection between the discharge line and the end of the heat exchanger.

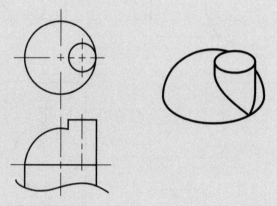

Figure 7.60

21. Figure 7.61 shows a spherical tank supported by sloping legs. There are four legs spaced at 90° (only one is shown). The diameter of the sphere is 2000 mm and the support is 500 mm diameter. The center of the support is offset 500 mm from the centerline of the sphere and the support has a slope of 75°. Draw the line of intersection between the support and the tank.

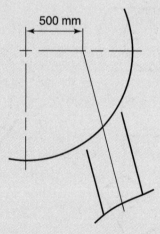

500 mm

Figure 7.61

Now that you have learned the concept of intersection, let's move on to look at development.

Development

If two pipes are to be joined, the end of one, and sometimes both, must be shaped so they meet. Some procedure must be used to mark the line of intersection (or what will be the line of intersection). A two-dimensional pattern is made and wrapped around the end of the pipe to mark the cut. The process of making a two-dimensional pattern that can be rolled or folded into a three-dimensional shape is called **development**.

You have seen many applications of developed objects, although you probably did not think of them this way. A cereal box is made by folding a flat piece of cardboard. The cardboard is cut so that the final shape is formed after folding (see Figure 7.62).

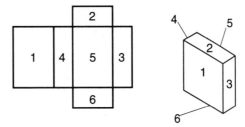

Figure 7.62 Pattern for a box

Large diameter pipe is made from a flat piece of steel by rolling and welding the seam. Mailing tubes are made by rolling a flat piece of cardboard.

A line of intersection must be found before a surface can be developed. This process has been discussed; however, when a development is required, surface lines must be equally spaced. Equal spacing is important only if development is required.

Development of a Cylinder

A cylinder is made by joining the edges of a rectangle, as shown in Figure 7.63. The lines drawn parallel to the axis at the seam are called **generator lines**. Two other generator lines are shown on the flat pattern.

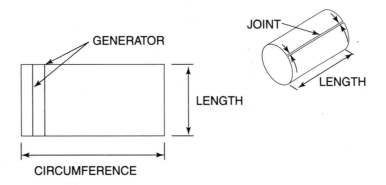

Figure 7.63 Development of a cylinder

If a specific length and diameter are required, as is usually the case, the material must be the right size. One side must equal the length and the other, the circumference.

The development of a cylinder with square ends (ends that are perpendicular to the axis) as shown in Figure 7.63 is simple. All generator lines are the same length. If one end is at an angle which is not 90°, all generator lines will not be the same length and the pattern is not so simple.

Example 7.9

Develop a pattern for the end of the cylinder shown in Figure 7.64 so that it can be cut to meet the wall. An auxiliary view showing the axis of the cylinder as a point view is shown in Figure 7.64.

A pattern for the end of this cylinder can be developed by starting with a rectangle, as in Figure 7.63, and marking the length of each generator line so that when rolled it will produce the cylinder in Figure 7.63. The problem is reduced to finding the length of a line (actually several lines). The first three steps are similar to those used previously to find the line of intersection, except that generator lines must be equally spaced.

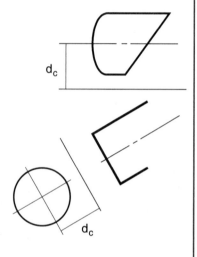

Figure 7.64
Cylinder with end cut at an angle

1. Locate equally spaced generator lines on the surface of the cylinder. This cannot be done in the plan view. Generator lines can only be positioned in a view showing the true shape of the cylinder (i.e., a circle). They will appear as points in this view. Generator lines have been located around the circumference by spacing them at equal angles (see Figure 7.65). Thirty degrees is a good spacing if the cylinder is small (or has been drawn to scale). Generator lines are numbered for reference. The numbers usually start at the seam (where the two edges are joined). Generator lines have been projected to the front view in Figure 7.65.

2. Locate these generator lines in the plan view to find where each line intersects the wall. They are located in the plan view using the distance from the folding line in the auxiliary view. Figure 7.66 shows how this distance is transferred to the plan view.

3. Project all intersection points from the plan view to the corresponding line in the front view and join them with a smooth curve to get the line of intersection. The line of intersection is shown in Figure 7.67. An arbitrary datum has been added. The length of each generator line will be measured from this datum, which can be in any convenient location. It helps to number generator lines.

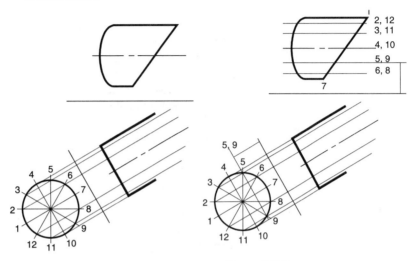

Figure 7.65
Locating equally spaced generator lines

Figure 7.66
Distance transferred to plan view

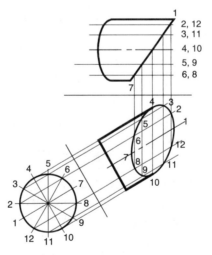

Figure 7.67 Line of intersection

4. Lay out the circumference (πd). This is shown as line AB in Figure 7.68. Generator lines must be located along the circumference and the length of each laid out.

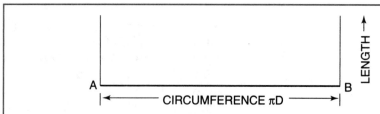

Figure 7.68 Circumference laid out

5. Divide the circumference into 12 equal parts so that generator lines can be located at equal intervals. You can calculate the circumference and divide it by 12, but it will probably not divide evenly, making it difficult to divide the line accurately. There is an easy way to divide the line into 12 equal parts. The process is shown in Figure 7.69.

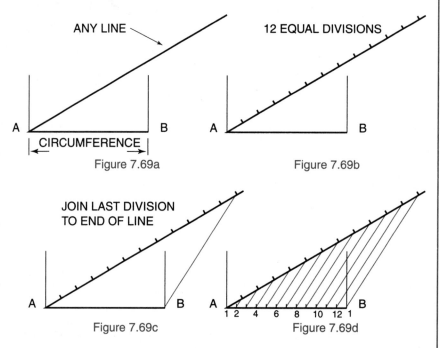

Figure 7.69 Dividing a line into equal parts

a. Draw a line, at any angle and any length (make it as long as possible), from one end of **AB** (Figure 7.69a). Small angles do not work well for this. Use an angle of about 30°. There is no need to measure it.

b. Mark off 12 equal lengths along this line (Figure 7.69b). Use any convenient interval, but it should not be too small.

c. Draw a line from the 12th division to the end of **AB** (Figure 7.69c).

d. Draw lines parallel to the line drawn in Figure 7.69c from each of the measured divisions. The circumference is now divided into 12 equal divisions (Figure 7.69d).

Figure 7.70 shows the circumference divided into 12 parts using this method. Generator lines have been drawn at each division. The length of each generator line must now be laid out.

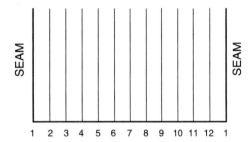

Figure 7.70 Circumference divided into equal divisions

6. Mark off the length of each generator line on the pattern. Length, measured from the arbitrary datum, is taken from the front view. The lengths of generator lines 5 and 12 are shown on the front view. Figure 7.71 shows the length of all generator lines on the pattern.

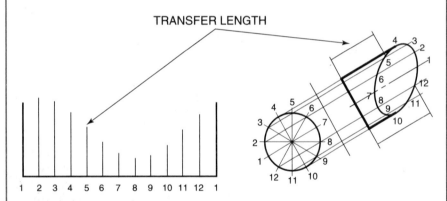

Figure 7.71 Generator line length laid off on pattern

7. Draw a smooth curve through the end of each generator line to finish the pattern (see Figure 7.72).

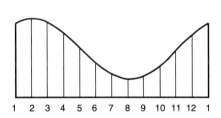

Figure 7.72 Development of end of cylinder

Development of Two Curved Surfaces

When two cylinders (or other curved surfaces) meet, development is done in the same way as for the intersection with a plane surface. The circumference is first divided into equal divisions and the line of intersection is found. The circumference is laid out and the length of each generator line determined.

Example 7.10

Figure 7.73 shows a joint in a frame made from large diameter pipe. A pattern must be made so the vertical pipe can be cut to fit the sloping pipe.

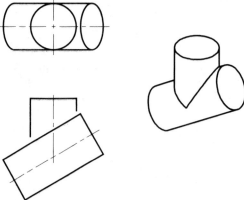

Figure 7.73   Intersection of two cylinders

You will create intersecting lines on the surface of both cylinders. The point where the lines intersect defines the line of intersection. The pattern can then be made. Figure 7.74 shows how corresponding lines will be seen on both surfaces.

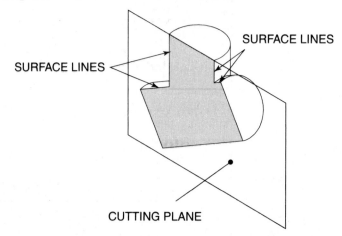

SURFACE LINES

SURFACE LINES

CUTTING PLANE

Figure 7.74   Using a cutting plane to create lines on the surface

1. Draw a plan view and divide the circumference of the vertical pipe into equal parts to give equally spaced generator lines. The point views of these generator lines are numbered in Figure 7.75.

2. Put cutting planes through the generator line locations. Cutting planes are seen in edge view in the plan view and are numbered to correspond to the generator lines (see Figure 7.76).

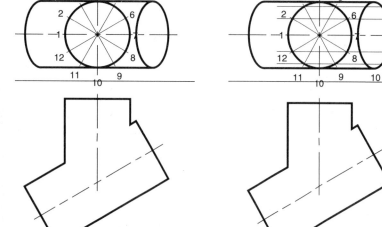

Figure 7.75
Generator lines found

Figure 7.76
Locating cutting planes at generator line locations

3. Draw an auxiliary view showing the axis of the sloping pipe as a point. Locate the cutting planes in this view using the distance from the folding line in the plan view. This creates lines on the sloping pipe that correspond to those on the vertical pipe. They are seen as points in this auxiliary view (Figure 7.77).

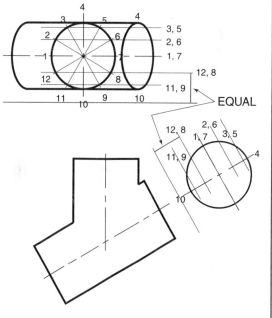

Figure 7.77   Creating corresponding lines on the sloping pipe

4. Project the surface lines on the sloping pipe to the front view and the generator lines on the vertical cylinder from the plan and identify the intersection points of corresponding lines (see Figure 7.78).

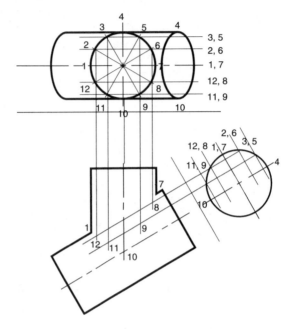

Figure 7.78 Intersection points found

5. Draw a smooth curve through these points to show the line of intersection (see Figure 7.79).

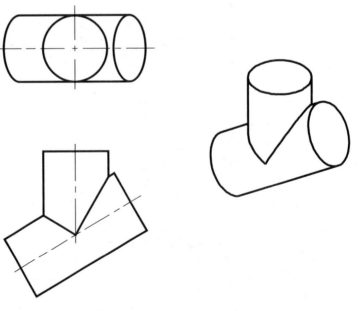

Figure 7.79 Line of intersection

6. Lay out the circumference of the vertical pipe and divide it into 12 parts, corresponding to the generator lines (see Figure 7.80).

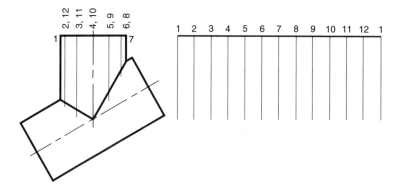

Figure 7.80   Generator lines located on the pattern

7. Mark off the length of each generator line on the pattern. The length of these lines is seen in the front view and is measured from an arbitrary datum. The length of each generator line is projected from the front view (see Figure 7.81).

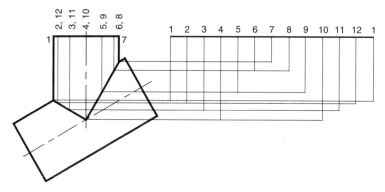

Figure 7.81   Generator line lengths laid out on pattern

8. Join the ends of the generator lines with a smooth curve (see Figure 7.82).

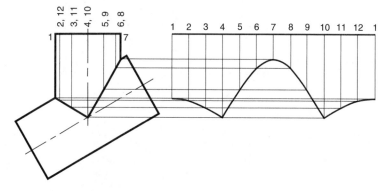

Figure 7.82   Completed pattern for vertical pipe

Now that you know a little more about development, try the following problems.

Problems

22. Figure 7.83 shows a tank with hemispherical ends filled with a horizontal line 600 mm diameter. The tank diameter is 1500 mm. The edge of the inlet line is on the centerline of the tank. Develop a pattern for the end of the inlet line so that it can be welded to the tank.

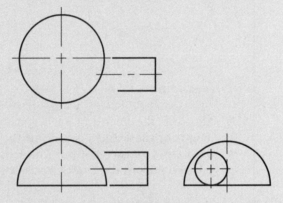

Figure 7.83

23. A front view of a 90° elbow supported by a cylindrical support is shown in Figure 7.84. The centerlines of the elbow and the support are in the same vertical plane. The cross-section of the elbow is circular, with a diameter of 400 mm. The diameter of the support is 300 mm. Develop a pattern for the support so that it can be welded to the elbow.

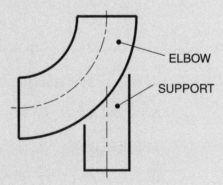

ELBOW

SUPPORT

Figure 7.84

24. Figure 7.85 shows the end of a cylindrical tank, diameter 600 mm, with a hemispherical end. There is a 360 mm discharge line at the top. The discharge line is in the plane of the centerline of the tank and slopes at 40°. Develop a pattern for the discharge line.

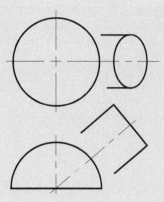

Figure 7.85

25. Figure 7.86 shows a 500 mm diameter cylinder supporting a 1000 mm diameter column. The centerlines of the two cylinders are in the same plane. The support makes an angle of 45° with the ground. Develop a pattern for the support. Both ends must be developed and the pattern must be the correct length. Locate the seam on the underside of the support. Choose a datum in the middle of the support and measure the length to the line of intersection to the top (at the column) and the bottom (at the ground). The scale is 1:20.

26. Plan and elevation views of a portion of a spherical tank supported by vertical legs are shown in Figure 7.87. Supports are spaced around the tank at 45° (only one is shown). The tank diameter is 8 m, and the support diameter is 2 m. Develop a pattern for the top of the support so that it can be welded to the tank. The scale is 1:50.

27. Figure 7.88 shows a 400 mm diameter feed line entering a 1000 mm diameter tank. The line enters the tank at an angle of 30° to the horizontal and is tangent to the side of the tank. Develop a pattern for the end of the feed line. The scale is 1:10.

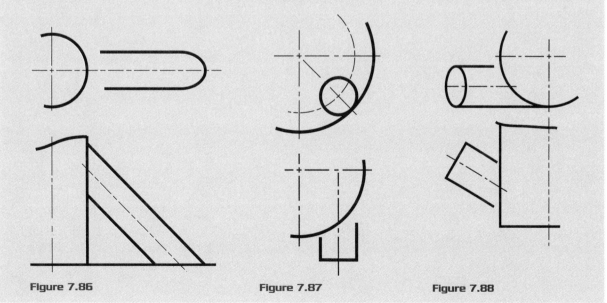

Figure 7.86   **Figure 7.87**   **Figure 7.88**

28. Figure 7.89 shows a line entering a torus. A torus is a moulding in the shape of doughnut. The diameters of the torus and the line are 400 mm. The centerlines of the line and the torus are in the same horizontal plane. Develop a pattern for the end of the line.

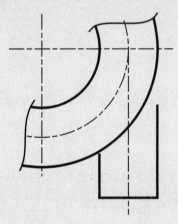

Figure 7.89

29. A cardboard box, 300 mm high, 200 mm wide, and 50 mm deep, is to be made from one piece of cardboard. Draw a pattern for this box so that it can be folded into the required shape. Allow 15 mm tabs so the seams can be glued. Dimension the pattern.

Chapter 8
Presenting Technical Information

The Importance of Being Understood

Your writing must not only be possible to understand, it must also be impossible to misunderstand.

If you ask engineering managers what they want in a new graduate, they will tell you that they want engineers who can communicate effectively with other engineers and clients.

Engineers do not spend all of their time in front of computers creating drawings and doing calculations. In fact, they spend only a small portion of their time drawing and calculating. So, how do engineers spend their time?

Engineers spend 60 to 70 percent of their time communicating: talking, writing, and listening. Ideas and problem solutions are of little value unless they can be communicated effectively to others. Records must be kept, and work must be documented. Information must be collected and questions asked. Documentation, even for a small project, can reach hundreds of pages. Whether information is stored electronically or on paper, someone must write it, and the engineer is often the only person who understands a project well enough to document it.

Information is no use unless it can be understood. If a prospective client (or your supervisor) does not understand what you are trying to say, you will probably not get a job (or keep one). A request for information that is not understood is unlikely to result in correct information. Records are of no use if they cannot be understood. Time and money are wasted if readers cannot understand what you are trying to tell them. Think about it. How many times have you read something twice because you did not understand it the first time? If your firm was hired to write a report and the client did not understand it, would the client hire your firm again? Unlikely. If a person reading your work must commit funds for a project and does not fully understand what you wrote, what are the chances of getting the funding? Slim.

In a nutshell, if you want to advance in your career and your supervisors do not understand what you are doing, good luck!

Assembling the Right Information

It is standard practice to document and file all information pertaining to an engineering project. To ensure that you gather all of the information you need, you may want to use the *Who? What? When? Why? How? and Where? process*. With this process, you ask yourself questions such as:

Process Question	Description
Who?	• Who was involved in the project? • Who made the decisions?
What?	• What did the project entail? • What was said? • What decisions were made? • What deadlines (timelines) were established? • What estimates were made? • What assumptions were made? • What action was taken?
When?	• When were the decisions made? • When was the project due to be completed? • When did the meeting take place?
Why?	• Why were the decisions made? • Why was the project delayed? • Why did the rafters not meet properly?
How?	• How were the decisions arrived at? • How was the problem resolved?
Where?	• Where is the project located?

By asking such questions you can start to build a complete picture of a project. Always keep in your mind that you may not be around to answer questions, so your documentation must reflect the five Cs of writing.

Using the Five Cs of Writing

1. Clear	Ensure that what you are writing is going to be interpreted correctly by the reader; e.g., place the word "only" in the correct position; keep the subject and verb together, etc.
2. Concise	Ensure that your communication is not wordy
3. Correct	Ensure that your communication is accurate as to factual content (names, locations, sums of money, etc.), spelling, grammar, etc.
4. Concrete	Ensure that your statements are specific; e.g., "The P95 valve was not working on April 9" as opposed to "The valve was not working."
5. Courteous	Ensure that all communications are polite, whatever you may feel. Choose words that will not offend the reader.

Keep in mind that records are valuable in preventing and resolving misunderstandings. Documents are often used years after a project is complete.

It is common practice for engineers to keep their own files of documents and correspondence relating to each project they work on. If well-written records are kept, there is no need for the engineer to have to remember what was said or done (an impossible task).

What Do Engineers Write?

Engineers write everything from letters to instruction manuals. The most common written communications are memorandums, letters, reports, specifications, standards, and codes.

Memorandums

A **memorandum**, often called a **memo**, is an informal, internal communication that you can send to one or more persons or simply file as a record. It is one of the most used methods of written communication. Figure 8.1 illustrates a standard memorandum format.

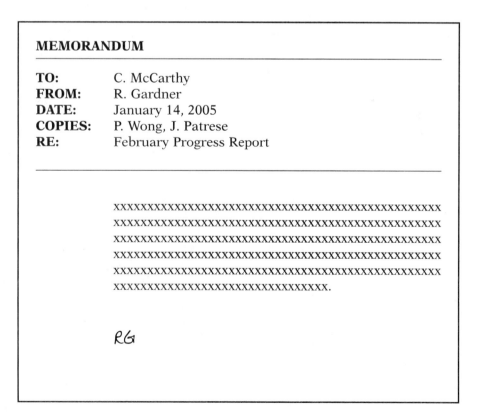

MEMORANDUM

TO:	C. McCarthy
FROM:	R. Gardner
DATE:	January 14, 2005
COPIES:	P. Wong, J. Patrese
RE:	February Progress Report

XXXXXXXXXXXXXXXXXXXXXXXXXXXXXXXXXXXXXXXXXXXXXXXXXXXX
XXXXXXXXXXXXXXXXXXXXXXXXXXXXXXXXXXXXXXXXXXXXXXXXXXXX
XXXXXXXXXXXXXXXXXXXXXXXXXXXXXXXXXXXXXXXXXXXXXXXXXXXX
XXXXXXXXXXXXXXXXXXXXXXXXXXXXXXXXXXXXXXXXXXXXXXXXXXXX
XXXXXXXXXXXXXXXXXXXXXXXXXXXXXXXXXXXXXXXXXXXXXXXXXXXX
XXXXXXXXXXXXXXXXXXXXXXXXXXXXXXXX.

RG

Figure 8.1   Sample memorandum format

The format of a memorandum is relatively standard. The guide words at the top of the memorandum are usually TO:, FROM:, DATE:, COPIES:, and SUBJECT: or RE: (meaning "regarding"). They may be organized in any order, though the subject or re: line should be the last item.

Many organizations have pre-printed forms. If your firm does not, ask one of the clerical staff in your office whether there is a memorandum template on the computer system. If not, use a template from your word processing software.

To save time, many organizations use initials in memorandums (instead of full names); for instance:

MEMORANDUM

TO: CMcC
FROM: RG

There is no restriction on the length of memos, but normally they are short.

While there is no need to sign memorandums, some writers like to initial them (see Figure 8.1).

Letters

Letters are formal communications between two or more people and are signed by the person who wrote them. The format of a business letter is standardized, although there are variations. Business letters will be dealt with in more detail later in this chapter.

Reports

Reports are written for a variety of reasons but the main ones are:

- to present the results of an investigation
- to present information on an event
- to describe a piece of equipment
- to make recommendations to a client on the solution to a problem.

All reports have a similar format. They may vary in length from a few pages to several thousand pages.

Engineering reports will be looked at in more detail later in this chapter.

Manuals

Manuals are written to tell someone how to do something. An assembly manual, for example, tells a person how to put something together. If steps are missing, the item will either be assembled incorrectly, or not at all.

Maintenance or instruction manuals that cannot be understood, or which have missing instructions, are not only frustrating but can also have serious legal and safety consequences.

Proposals

Engineering firms submit **proposals** to prospective clients to indicate how they would do the work, how long the work would take, and how much it would cost.

An unclear or ambiguous proposal is not likely to be accepted. If your proposal is not accepted, it means loss of work. An engineering company that does not get jobs will not survive.

A proposal that has spelling mistakes, poor grammar, and missing words will not inspire confidence in the firm that submitted it.

Specifications

Specifications are written to define how something must be done or how it must perform. If you want bids on a heat exchanger, for example, you would write a specification giving the quantity of heat to be exchanged, the fluids (water, engine oil, steam, etc.), temperatures, and other information, and ask heat exchanger manufacturers to quote on the project.

A specification on a building would outline the details of the building and what the contractors must do.

Standards

There are **standards** for almost everything in engineering; for example, there are standards that define the content and strength of metals and alloys. The American Society for Testing and Materials (ASTM) publish standards for testing materials and for the materials themselves. This ensures that a designer can specify a material and know how much load it can carry.

The Society of Automotive Engineers (SAE) set standards for the automotive industry.

Codes

Codes are similar to standards. They define construction and performance criteria for the protection of the general public. For example, the American Society of Mechanical Engineers (ASME), Boiler and Pressure

Vessel Code specifies how vessels operating at high pressures and temperatures must be built and tested. Pressure vessels that do not meet this code will not be licensed to operate by local authorities.

National and local building codes define how public buildings must be built.

The Writing Process

Writing well is not an easy task. It requires the same thought and preparation as other engineering tasks. It also takes time. Do not start the night before a written communication is due and expect to do a good job. Even experienced writers revise their work several times.

Writing starts with you, the writer, knowing what you want to say and why you want to say it. This seems obvious, but beginning writers often start with no clear idea of what they want to say and this is reflected in what they write. The result is a communication that is confusing and difficult to read. This is not surprising: Good writing is not the product of a confused mind. You must be clear and specific about what you want to say. (Remember those five Cs of writing mentioned earlier.)

You may find the following six-step process to be useful:

Step 1: Identify Your Objective
Step 2: Prepare an Outline
Step 3: Identify Your Audience
Step 4: Prepare a First Draft
Step 5: Edit Your Communication
Step 6: Review the Final Product.

Step 1: Identify Your Objective

Start by defining the objective of your communication. What is your reason for writing?

- Do you want to explain something?
- Do you want to recommend something?
- Do you want to describe something?
- Do you want to ask for information?
- Do you want to give information?

Then ask yourself what you must tell or ask the reader to do in order to meet this objective. For instance, if you need to explain something, what does the reader need to know? If you are making recommendations, what information must you provide to support your recommendations? What would you like the reader to do after reading your work? If the reader must make a decision based on your work, what information must the reader have to make the decision? Write down the answers to these and other questions like them, in as much detail as possible. More items can be added later if required. To make life easier for yourself, use a word processor.

If you can come up with a clear, precise, written statement of your reasons for writing, you will make your work easier. Read this statement from time to time to stay focused on your objective.

Step 2: Prepare an Outline

When you have a clear idea of your objective and what you want to say, you can plan how you want to say it. To do this, create an **outline** using a word processor. Figure 8.2 illustrates a sample outline.

Introduction
Project Specification
Project Objectives
Preliminary Design Solutions
    Gravity Launcher
    Pulley Launcher
Project Design I
    Launcher Description
    Testing and Adjustments
    Calculations and Assumptions
    Performance Evaluation
    Cost
Project Design II
    Launcher Description
    Testing and Adjustments
    Calculations and Assumptions
    Performance Evaluation
    Cost
Conclusion

Figure 8.2   Sample outline

The following explains the process of preparing a draft outline:

- The first thing on your outline will normally be an **introduction** or **background**. How detailed this introduction or background is depends upon what the reader knows about the subject, but a short introduction is a good way to lead into your topic.
- Next, list the main topics you want to cover. Don't worry, if you leave something out, you can always add it later. If you are using a word processor, this will be an easy task.
- Then, look at each of the main topics and make notes under each heading. These notes can be sub-topics, words, or phrases. It does not matter. Notice in Figure 8.2, that the sub-topics are indented. This helps to keep you organized.

This first, draft outline will be a "skeleton" or table of contents of what you want to write. It will give you an idea of how you can organize the material for maximum impact. If you are using a word processor, you can move items around on your outline until you are happy with the "flow."

We'll look at various organizational methods a little later in this chapter.

Step 3: Identify Your Audience

Engineering communications are intended to be read. There is not much point in writing something that is never going to be read or understood. Even records, written and filed, are intended to be read at some time or another. Why else would records be kept?

When you are writing, the reader and the needs of the reader must be kept in mind at all times. If the reader, or readers (there is often more than one) do not understand what you have written, you have wasted both the reader's time and your own.

You will usually know who the readers will be (or at least their background) and you must write with this in mind. A reply to a question from a Grade 9 student would not be written in the same way as it would be if it were to another engineer (at least, it shouldn't be).

It is reasonable to assume that most of the readers of an engineering report have some technical background, but it may be in another field of engineering. Do not assume that the readers will have the same background as yourself. Your readers should not have to guess at the meaning of any terms. A footnote at the bottom of the page, an endnote at the end of a section, or a glossary of terms can be used to help readers, if necessary. Remember, readers will try to follow your meaning but eventually, even the most persistent, give up. If the readers are people who can affect your future, you will not do yourself any favors by using unusual or complicated terms. Make it easy for the readers. Make sure that they remember you for clarity, not confusion.

Step 4: Prepare a First Draft

You are now ready to write the first draft. (Remember what I said about not leaving it to the day before it is due?)

The objective of the first draft is to get your ideas on paper, not to create the finished work. Do not worry too much about how you express yourself. Use your outline. Write short, concise paragraphs under each heading. This will help you to keep focused. Prepare the first draft as quickly as you can (following your outline). There is always time to work on the sentence structure and flow in the second and subsequent drafts.

When you have finished your first draft, use the grammar- and spell-checker on your word processor to clean up your work. Print a copy of your first draft, but put it aside until the next day.

Step 5: Edit Your Communication

The next day, read your first draft. It helps to read your work out loud. Remember, you will write several drafts of an important communication before it is sent out. Even experienced writers do not consider the first draft a finished product.

As you read, edit your work for flow, extra words, slang and idioms, gender-inclusive language, and spelling and grammar.

Flow

When a document flows, it is easy to read, so edit your work so that one idea flows naturally into another. Make sure that each paragraph in your communication expresses one main idea.

Extra Words

Edit your work to remove redundant words. Quantity does not equal quality.

The average first draft can be shortened by one-third by removing excess words. But, you may ask, how can I remove words and improve my work? How can I identify excess words? The easiest answer to these questions is to remove words and see what happens. Here is an example:

The following sentence contains frequently used words:

> *However, it should be noted that extra words are a waste of time to read and write.*

Removing redundant words reduces this to:

> *Extra words are a waste of time to read and write.*

The number of words has been reduced by 35 percent but the meaning has not changed.

Make every word count. If you can say it in five words, don't use ten. Your readers are busy. Don't waste their time by making them read unnecessary words.

Slang and Idioms

Avoid slang such as: "There's no way that this project will fly" and idioms such as: "I think we need to emphasize that they must keep their eye on the ball." While you may understand what you are saying, remember that engineers work in a global environment. What may seem everyday, understandable language to you, may be totally "foreign" to the overseas reader.

Gender-inclusive Language

As you are editing, watch for any stereotypic references; for instance, referring to an engineer as "he." If you reword the sentence into the plural, you can eliminate this problem; e.g., "Engineers ......they ....."

Spelling and Grammar

A letter or report full of spelling or grammatical errors does not inspire confidence in the writer. If you are sloppy with your spelling and grammar, what about your engineering work? Poor quality work reflects badly on both the writer and the firm. You cannot use the excuse, "the typist made some spelling errors." The person responsible for finding errors is the person who sends the communication.

Ah, but you have a word processing program with a spell-checker so all you have to do is run a spell-check, right? Think again. What would a spell-checker do with the following sentence?

Hoe wood you two this problem?

There are no spelling mistakes, but the sentence makes no sense. These are all common words that are in the "dictionary" of your word processor. Errors like this are found by proofreading your work, not by a spell-checker. A spell-checker may find errors if the words are misspelled but not if the wrong word is used. If you received a letter with errors like the above, from someone doing engineering work for you, what would your impression be?

Use the spell-checker by all means (it is surprising how many people don't), but do not rely on it completely. The same thing applies to the grammar-checker. Your best solution is to proofread your work *several times*. Buy a good dictionary and style manual and use them if you are in doubt.

Proofreading

As you proofread your communication, make sure that no words are missing and that you have used the correct words. Ensure that your message is clear. Remember that communications written in the active voice will be more concise and "punchy." When your word processor identifies passive voice sentences, look at the suggestions provided in the grammar-checker. You do not have to follow the suggestions, but sometimes they are useful.

The following proofreading hints may be of assistance to you:

Proofread for:	Hints
Spelling	• Check for spelling errors, even if a spell-checker has been used. • Check for out-of-context words. A spell-checker will not find words used out of context; e.g., "there" instead of "their." • Check for repeated words. Some spell-checkers will not find repeated words; e.g., "This is is the radius." • Check names of people and places carefully.
Grammatical Errors	• Look for incorrect changes in verb tenses; e.g., changing from the present tense to the past tense. • Ensure that subjects and verbs match; e.g., check that singular nouns have singular verbs. • Check that every sentence is, in fact, a sentence. Each sentence must contain a verb. • Check parallelism; e.g., bulleted items (such as in this list) start with the same part of speech (verb); that words in a series match.
Accuracy of Facts	• Check numbers. • Check names and dates. • Check page and section references; e.g., if the report text states that something is on page 14, check that it is really there.
Hyphenation	• Ensure that the last word of a paragraph or page is not hyphenated. • Check that words are correctly hyphenated. Use a dictionary or word division reference book if you are uncertain. • Ensure that there are no more than three consecutive lines ending with a hyphen. If there are, edit the text.
Punctuation	• Check the spacing after commas, periods, colons, semi-colons, etc. • Check that all question marks are inserted. • Check that all commas, semi-colons, parentheses, brackets, apostrophes, and quotation marks are correctly placed. • Ensure that underscores do not extend beyond the words to be underlined. • Ensure that parentheses or brackets around bold or italicized text are not bold or italicized.
Page Layout and Alignment	• Ensure that paragraphs are indented the same distance (if an indented style used). • Check that all columns align at the bottom of the page if you are using a multi-column format. • Check that all figures in a column are aligned. Dollar amounts should align on the decimal point. • Check that all margin settings are the same for each page. • Ensure that all pages have the same format. • Check all pages for balance.

	• Ensure that there are no "widows" or "orphans": one line of a paragraph left at the bottom of a page or the top of a page.
Consistency	• Check that headings of the same level use the same font and format (use styles).
	• Check that all captions are labeled in the same manner.
	• Check that all sections and units are referenced in the same manner; e.g., Section 4 in all places, not Section IV in one place and Section 4 in another.
	• Check that all tables are formatted the same; e.g., that all have centered column headings or all have left-aligned column headings.
	• Check numbered items or lists for missing numbers.
	• Check headers and footers to ensure that they relate to the material on the page, section of the report, or the report itself.
	• Ensure that the page numbers are in order and that none are missing.
Graphics	• Ensure that any graphics are referenced in the text.
	• Ensure that the captions on graphics match the graphics themselves.
	• Check that the graphics are there.

[Adapted from Porozny, G. H. J. *Desktop Publishing Design Basics and Applications*. Toronto: Copp Clark Pitman Ltd., 1993]

Keep editing your communication until it is clear, concise, correct, concrete, and courteous; however, avoid the perfectionist trap where you keep editing without any improvement to quality.

Step 6: Review the Final Product

When you have thoroughly proofread your final draft, print it on good quality, white bond paper (or letterhead, if the communication is a letter).
Review the final product:

- Is there plenty of white space on the page or is the text cramped?
- Are the top, bottom, and side margins too wide or too narrow?
- Are there any messy spots or fingerprints on the printout?
- Is the printout clear or do you need to add toner to your laser printer?
- Does this communication look as if it has been prepared by a professional organization?
- Would you like to receive this communication?

If you are satisfied with the quality of your communication, sign it (if appropriate).

Business Letters

The **business letter** is a formal communication between two people. It is the most common form of communication in business.

Business letters all share a similar format. A standard format enables the reader to find out who the letter is from, who it is intended for, and what it is about without reading the entire letter. If the subject of the letter is such that it should be passed to others for action, this can be done without taking the time to read the letter.

In addition to a common format, most business letters are short.

The most common business letter format is the block style illustrated in Figure 8.3.

Advanced Analysis Group Inc.
1421 Albert Street
Vancouver, British Columbia, Canada
V7R 3E7   **Letterhead**
Phone: (604) 425-9871
Fax: (604) 425-1327
e-mail: advanced@sympatico.bc.ca

Our File 04-A56

July 20, 2006

Ms. Jane P. Yee, President
International Equipment Supply Limited   **Inside Address**
1587 Vernon Street
Vancouver, BC V6T 1Z9

Dear Ms. Yee: **Salutation**

Re: X-14 Diesel Engine Oil Cooler Investigation   **Subject Line**

We enclose our report, "Failure of the X-14
Diesel Engine Oil Cooler," which you requested
on June 15, 2006. Our investigation showed that   **Body**
the cooler failed because of poor installation. The
events leading to this failure are detailed, and suggestions are made to prevent it happening again.

We will be pleased to discuss this with you at your
convenience.
                                                              Complimentary
Yours truly,   **Closing**

ADVANCED ANALYSIS GROUP INC.

W. A. Baxter   **Signature Block**

Wm. A. Baxter, Ph.D., P.Eng.
President

Figure 8.3   Block style letter

Letterhead

The letterhead, which is normally preprinted, gives the name of the company sending the letter, their address, their telephone and fax numbers, and often their e-mail address or web site. Some companies list the company name in the letterhead and the remaining information at the bottom of the page.

If a letter is more than one page long, letterhead is used only for the first page. Continuation paper is used for the second and subsequent pages. Figure 8.4 shows a sample continuation page.

Advanced Analysis Group Inc.

Ms. Jane P. Yee
Page 2
July 20, 2006

events leading to this failure are detailed ...

Figure 8.4   Continuation page

Inside Address

The inside address gives the name of the person receiving the letter, their title (if any), and their complete address.

Salutation

If you are writing to a specific person, as in Figure 8.3, then the name of the person is used in the salutation.

If you are not writing to a specific person, you can use a general salutation such as:

Dear Sir:

Dear Sirs:

Dear Madam:

Subject Line

The subject line tells the reader what the letter is about. It should be specific. If the reader has no interest in the subject, there is no need to read further. The letter can be forwarded to the appropriate person for action.

Body

The body of a letter is what you want to say. Remember, keep it short and precise. Ensure that it has a logical flow to it: Introduction, middle, and ending.

Complimentary Closing

Keep the complimentary closing simple. "Yours truly" and "Sincerely yours" are commonly used.

Signature Block

The signature block contains your name and title (if appropriate). You sign in the space between the company name and your own.

There are variations on this format and some companies have their own format.

Engineering Reports

Reports are required at all stages of a project. They may be formal or informal but, at some stage, a formal report must be prepared.

Format

An engineering report can vary from a one-page letter to thousands of pages. The longer the report, the more important it is that the reader be able to find information quickly. A report need not be read from start to finish, and usually is not. For example, one reader may read only the conclusions, while another may be interested only in the costs.

A well-formatted report is divided into short sections and subsections to make it easy to refer to and easy to read.

The following are the main sections of a report.

- Cover
- Title page
- Abstract
- Table of contents
- List of figures (if necessary)
- List of tables (if necessary)
- Introduction
- Discussion
- Conclusions (if necessary)
- Recommendations (if necessary)
- Appendix(ces)

- References (if necessary)
- Bibliography (if necessary).

The purpose of your report will determine which sections you include; however, all reports have a cover and title page.

The following table briefly outlines the various parts of a report.

Report Section	Description
Cover	- The cover of a report usually has nothing on it other than the report title. The title must be specific and give the reader as much information as possible in about eight words or less. - Many companies have standard report covers with a simple design and company logo on them. The title is added to the cover. - Many report covers are printed on colored light-weight cardboard or heavyweight paper.

Title Page	- The title page format varies from company to company, however, it will have on it:
	  – the report title (same as on the cover)
	  – who wrote the report
	  – who the report was written for
	  – when the report was written.
	- The title page is page "i" of the report but the number is not printed on the page.

Abstract	- Sometimes called a **Summary**, the abstract is a "report-in-miniature." It contains an accurate summary of the topic, important results, and conclusions. - The abstract must be brief and concise. The reader must be able to understand the key elements of the report by reading the abstract. - The abstract should contain no reference to the report itself such as "this report contains…" nor should it contain any reference to sections, figures, or tables. - The abstract is normally placed immediately after the title page (as page ii), unless there is a preface. - Many people consider the abstract to be the most important part of a report. Consider writing it last.
Table of Contents	- The table of contents functions as an outline of the report. It lists all headings and major topics and indicates the pages on which they can be located. - The table of contents usually starts at page iii. - If you are using word processing software, you can generate a table of contents automatically provided that you have assigned styles to the headings in your report.

List of Figures	• The list of figures gives the figure number, its caption, and the page number where it can be found. • A short report does not require a list of figures.
List of Tables	• The list of tables gives the table number, its caption, and the page number where it can be found. • A short report does not require a list of tables.
Introduction	• The introduction states why the report is being written, what it is intended to achieve, the scope of the report, the plan of development, and the general conclusion or recommendations. • Some background on the project may be included, depending upon the intended reader; for instance, if the report is intended for someone who is familiar with the project, less background material is required. • The introduction should start on a right-hand page and be numbered (Page 1).

Discussion	• Sometimes called the **body** of the report, the discussion is normally the longest section of the report.
	• It presents facts, data, and arguments to support the conclusions.
	• All information is relevant.
	• All information is presented in an organized, logical manner. The following are some sample organizational methods:
	  – **Problem-analysis-solution** (presents the problem, analyzes it, and then proposes a solution)
	  – **Cause-and-effect** (presents the cause of the problem and what effects the problem will have; you may or may not want to include recommendations)
	  – **General-to-specific** (starts with a general statement and provides specific examples)
	  – **Simple-to-complex** (moves the reader from a simple concept through to a more complex one)
	  – **Question-and-answer** (starts with a question and then proceeds to answer it)
	  – **Chronological** (presents information in date order or as a sequence of events/steps)
	  – **Geographical** (divides the material accordingly to physical location).
	• Supporting information such as calculations (unless essential for reader understanding) is put into an appendix or series of appendices and referenced in the discussion; for example, "...see Appendix B-1."

Conclusions	• Conclusions must be based entirely on the material contained in the report. • New information, or information you knew

	before you wrote the report, is not a conclusion.
	• Put each conclusion into a separate paragraph. Consider numbering each conclusion for easy reference.
	• Some reports do not have conclusions; for example, a report that describes something, does not have conclusions.
Recommendations	• Recommendations must follow logically from the conclusions and be supported by the material in the discussion.
	• Recommendations must be clearly stated so that there are no misunderstandings.
Appendix(ces)	• An appendix or series of appendices contain supporting information that is not essential to the understanding of the report.
	• Each appendix deals with just one subject, and is assigned a letter; for example: – Appendix A: Calculations – Appendix B: Equipment – Appendix C: Physical Properties.
	• If you have appendices with related information, you may assign a letter and number; for example: – Appendix A-1: Materials List – Warehouse – Appendix A-2: Materials List – Production Facility I – Appendix A-3: Materials List – Production Facility II.
References	• The references section lists, in alphabetical order, all sources you have *referred to in your report*. Each entry includes the author(s), the title of the material, the publisher's city and province/state and publisher's name, and the date and year.
Bibliography	• The bibliography lists, in alphabetical order, all sources you have *used in writing your report* (which you may or may not have referred to in your report). Each entry includes the author(s), the title of the material, the publisher's city and province/state and publisher's name, and the date and year.

Presenting Your Information

How you present your report has an effect upon how it is received by the reader. A poorly formatted report, crammed onto "dog-eared" or finger-marked pages, or a computer presentation (such as PowerPoint®) with difficult-to-read visuals, is not going to be well received. While you may have the best proposal or solution to a problem, a poor presentation will not make a good first impression.

You need to combine the elements of your communication: text, draw-ings, graphs, photographs, etc. to give the best possible visual effect. You do not need to be a professional desktop publisher to create a visually pleasing document. The following are some tips from "the professionals" that you may find useful:

Text

Hints

- Use double spacing between lines of type. Use at least quadruple spac-ing if you are preparing a PowerPoint® presentation.
- Use easy-to-read fonts (Times New Roman, Arial, etc.).
- Consider using serif fonts (ones with little arms and legs on the letters) such as Times New Roman for body text because they are easier to read.
- Consider using sans serif fonts (ones without little arms and legs on the letters) such as Arial for captions, labels on figures, etc.
- Avoid using more than three font styles in your communication.
- Take careful advantage of font attributes such as size, bold, italics.
- Use a readable size of font (normally somewhere between 10 and 13). Use a larger font if you are preparing a PowerPoint® presentation because it normally has to be viewed at a distance.
- Use uppercase (capital letters) sparingly: only for labels, single words, or very short lines.
- Use a consistent paragraph format style (aligned left or justified).
- Use colored text and backgrounds with care when preparing a PowerPoint® presentation; e.g., blue text on a blue background will not work.

Compare the single-spaced text below the graph in Example 8.1 with the double-spaced text in Example 8.2. Notice how much easier Example 8.2 is to read. The additional **white space** between the lines of text helps readability and makes the page look "lighter."

Example 8.1

$K := slope(x, F)$ $K = 297.009 \ \dfrac{N}{m}$

$b := intercept(x, F)$ $b = -82.39 \quad N$

$y := K \bullet X + b$ (best fit line calculated from linear regression)

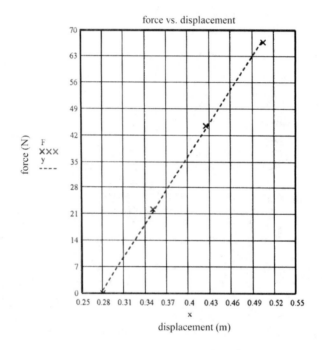

From the best fit line fo the date points we obtain a spring constant of 297.009 N/m. To calculate the energy stored in the spring we take the integral of the force over the displacement of the spring. In our launching device, the spring is stretched somewhat at the point of release, so our bounds of integration will be from the displacement of the spring at ball release, to the total displacement of the fully stretched spring.

Displacement of spring at ball release $x1 := \dfrac{(39.5 - 28)}{100}$

Total displacement of spring $x2 := \dfrac{(71 - 28)}{100}$

Mechanical Engineering 251 – Project II: Projectile Launcher

[Reprinted, with minor modifications, with the kind permission of Patty Lee, Cari Orbeck, Darren Rafferty, and Carola Veloso, Projectile Launcher Project Report, November 1998]

Example 8.2

Projectile Launcher   2

calculated that the piston could not generate enough velocity to accelerate the combined weight of the piston, cup, and squash ball. Even with more pressure, the piston would not be able to generate enough velocity to launch the ball ten meters. The cylinders are also only rated to 150 psi, so we do not want to be working in the upper ranges of the allowable pressures of the cylinder. A new idea incorporating some mechanical advantage would need to be considered for the launcher to be successful.

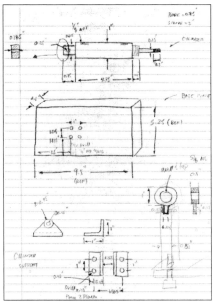

Figure 2: Initial experiment sketches

1.2 REVISED DESIGN

We still wanted to use the air cylinder and solenoid idea, since we saw how accurate it was, but needed some kind of leverage to increase the ball's initial velocity. One idea was to put a spring in the launcher cup so that it would compress as the piston shot out and then the spring would extend as the piston stopped. This would give the ball an extra push as it was released. This idea might have worked, but it would have been too hard to time the release of the spring to the stopping of the piston. Another idea we had was to have the piston shoot backwards and hit a lever that would catapult the ball towards the target. Figure 3 is a copy of our preliminary sketch. The cup used in this design turned out to have too high edges, resulting in a vacuum within the cup as the ball left it. To eliminate this, we cut away the back edge, doubling the distance of the launch.

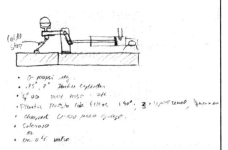

Figure 3: Preliminary Sketch

[Reprinted, with minor modifications, with the kind permission of Neil Allyn, Kurtis Guggenheimer, Alex Tielker, and Matei Ghelesel, Projectile Launcher Technical Report, 1998]

When preparing materials for a PowerPoint® presentation, make sure that you have plenty of "white space" or background color, and that your text is not crammed on the visual.

It does not matter what paragraph style you use. Both Examples 8.1 and 8.2 use an aligned left style (the left side of the text is blocked and the right side is ragged). This book uses a justified style (both left and right sides of the text are blocked). Be critical. If you select a justified style but your word processing software leaves too much space between words, use an aligned left style. Most word processing programs have various methods to rectify the spacing problem, but if you are not a word processing expert and do not have time to be, then use the simplest style you know that looks good.

It does not matter whether you indent the first line of your paragraphs or not. If you do, make sure that your indentations are consistent (1/2", 1.25 cm is normal).

Look again at Examples 8.1 and 8.2. Do you notice that the paragraph of text below the graph in Example 8.1 looks "heavy"? Do you realize that your eyes are drawn to this "heaviness" as opposed to the graph that is "light and airy"? Do you want this to happen? Usually, no.

Spend a few minutes looking at the text in this book. Ask yourself the following questions:

- Is the text easy to read?
- Are the fonts simple?
- Is a serif font used for body text and a sans serif font used for captions and labels?
- Are the font sizes and attributes used sparingly?
- Are capital letters (uppercase) used sparingly?
- Is there plenty of white space?

Single-page v. Double-page Spread

Before you start preparing a report, decide whether you are going to print on one side (single-page spread) or both sides of the page (double-page spread). It will make a difference as to the page set-up you select on your word processing program.

This book uses a double-page spread. As you open the book, there are two pages in front of you, aren't there?

Here are some hints for setting up a double-page spread report:

Hints

- Page numbers must alternate (odd page numbers – 1, 3, 5, etc. – on the right-hand pages, even page numbers – 2, 4, 6, etc. – on the left-hand pages).
- Page numbers must be on the outside of the alternating header or footer, unless you are using a centered page number style.
- Use an alternating page set-up so that the left margin on the left-hand page and the right margin on the right-hand page are the same width (often called *mirror margins*).
- Ensure that you allow a sufficient "gutter" (right margin on the left-hand page and left margin on the right-hand page) so that the report can be three-hole punched without damaging the text.
- Ensure that the "gutter" is wide enough so that the text does not disappear down the center of the two-page spread.

If you are using a one-page spread, you do not need to worry about alternating headers and footers (including page numbers) or gutters.

Graphics

Graphics comprise anything that is not text; for example, photographs, diagrams, drawings, sketches, maps, and graphs. They can be displayed in a number of different ways. Take a few minutes to look back at the graphics in Examples 8.1 and 8.2, as well as the ones in the following Examples 8.3, 8.4, and 8.5. Examples 8.4 and 8.5 illustrate engineering drawings and exploded assembly drawings included in an appendix to a report.

Example 8.3

Projectile Launcher   3

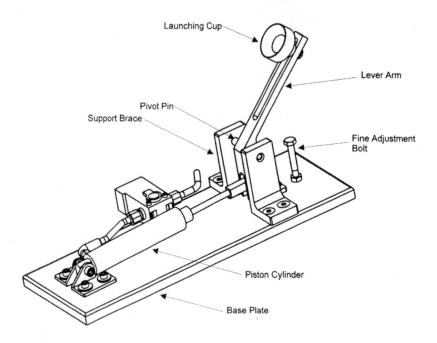

Figure 4: 3D representation

2.0 FINAL DESIGN

In the following section we will talk about our final design, detailing how and why it was made in certain ways. Figure 4 shows a 3D representation of our final design.

2.1 MANUFACTURING

We roughly calculated how much mechanical advantage we would need to increase the ball's velocity to the minimum necessary, and then decided on a length for this lever arm.

The first lever arm that we built was made out of thin-walled, square-sectioned aluminum bar-stock. We mitered two pieces of bar-stock at 22.5°, and then jigged and welded them together to make a piece with a 45° angle. We mounted the piston and the lever so that the distance ratio of the cup and the piston was about 3.5:1. We made our launcher

[Reprinted, with minor modifications, with the kind permission of Neil Allyn, Kurtis Guggenheimer, Alex Tielker, and Matei Ghelesel, Projectile Launcher Technical Report, 1998]

Example 8.4

Projectile Launcher   10

Appendix B: Drawings

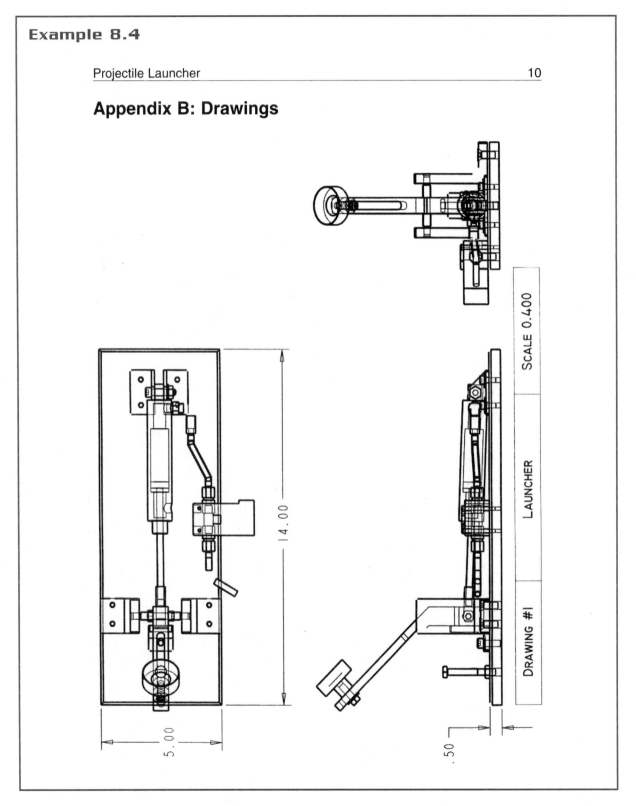

[Reprinted, with minor modifications, with the kind permission of Neil Allyn, Kurtis Guggenheimer, Alex Tielker, and Matei Ghelesel, Projectile Launcher Technical Report, 1998]

Example 8.5

Projectile Launcher   11

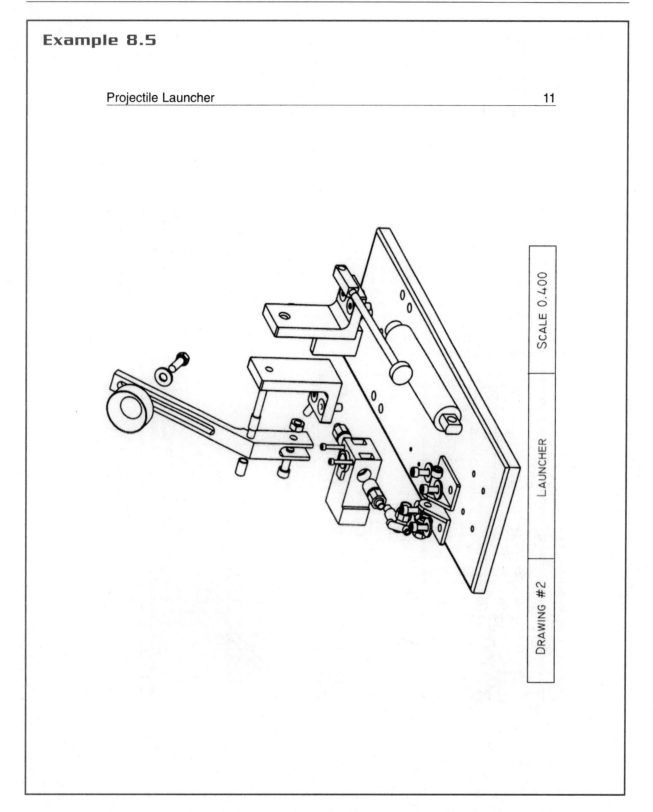

SCALE 0.400

LAUNCHER

DRAWING #2

[Reprinted, with minor modifications, with the kind permission of Neil Allyn, Kurtis Guggenheimer, Alex Tielker, and Matei Ghelesel, Projectile Launcher Technical Report, 1998]

The following are some hints for graphics presentation:

Hints

- Reference the graphic in the text; e.g., "Figure 3 illustrates ..."
- Place the graphic as close to the text reference as possible (always after or beside the text, never before it).
- Put the graphic number in the caption; e.g., Figure 3.
- Put a brief description in the caption; e.g., Figure 3: Profile of the mountain.
- Be consistent with capitalization and punctuation in the caption; e.g.,
  - Figure 3: Profile of the mountain
  - Figure 4 - Topographical Map. [Incorrect]
  - Figure 4: Topographical map [Correct].
- Center graphics horizontally on the page (see Example 8.1).
- Balance graphics diagonally on the page for an attractive page layout (see Example 8.2).
- Wrap text around graphics, where appropriate.
- Use appropriately sized graphics to balance with the text on the page (see Example 8.3).
- Place a graphic less than a third of a page deep at the top of a page and one more than a third of a page deep at the bottom of a page: This will ensure good page balance.
- Use landscape mode (sideways display) as opposed to portrait mode (vertical display) to fit detailed drawings onto the page (see Examples 8.4 and 8.5).
- Ensure all labels on graphics are easy to read and consistent as to size and capitalization (see Example 8.3).
- Save time by creating graphics in the simplest computer program. For instance the graph in Example 8.1 has been prepared in a spreadsheet and inserted into the report; the freehand sketches in Example 8.2 have been scanned and inserted into the report.
- Consider linking (as opposed to embedding) spreadsheets and graphs into your report, then, if you have to make changes to your spreadsheet or graph, the changes will automatically update in your report.
- Ensure that the values on the axes are appropriate so that the graph is not distorted, either horizontally or vertically. Keep in mind that the width of your report page may be less than your spreadsheet printout page. Graphs should be "open" and easy to read.
- Ensure that your pie charts contain no more than seven segments.
- Avoid cluttering graphs with unnecessary grid lines or labels.
- Avoid using a variety of bold patterns as fills on graphs or charts.
- Avoid cluttered graphics when preparing a PowerPoint® presentation. Labels may have to be read at a distance, so keep your reader in mind when selecting a label or diagram type size.

While it may seem that there are many things to think about when preparing communications, remember that the time will be well spent. Professional-looking communications project a professional image.

Problems

1. Write a proposal to a prospective client, who is not an engineer, describing your proposed design for one of the problems in Chapter 4. Illustrate your proposal with clear sketches or drawings.

2. Choose a common object and write a report on how it works. Select something that can be obtained easily at little or no cost (e.g., an old toaster, an old power tool), dismantle it, and, with the aid of clear sketches and drawings, describe the various components and how they work together. Find out the correct names for the various parts. Photographs are a good method of showing what the components look like and how they are assembled.

3. Write a set of instructions to explain how to program a VCR to perform a function, such as how to record at a specific time. Write your instructions, illustrated with sketches or drawings, so that someone who has little or no experience operating a VCR can perform the function.

4. Write a description of how some piece of equipment operates; e.g., a gasoline engine, an air compressor, a refrigerator, or a gas turbine. Ensure that your description can be understood by a Grade 9 student.

5. Write a set of instructions for conducting a detailed engineering search on the Internet. Write your instructions so that someone who has only a little knowledge of the Internet, or who is searching for something specific, can use them.

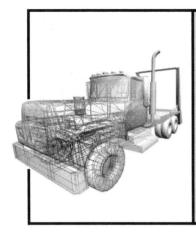

Appendix A

Vector Analysis

Vectors

A **vector** is a quantity that is defined by two things: magnitude and direction. **Force** is a vector quantity; it has magnitude and direction. It is important to know in which direction a force acts; for instance, if you want to lift something off the ground you do not push down on it. **Velocity** is another vector quantity; it has magnitude, speed (kilometres per hour), and direction. It can make a big difference if you are moving north instead of south.

We will deal with the graphical representation of vectors; however, vectors can be represented mathematically. The graphical representation of a vector quantity must indicate both the magnitude and the direction of the vector quantity, say, a force. Figure A.1 shows horizontal force acting on a body. The magnitude of the force is 50 N.

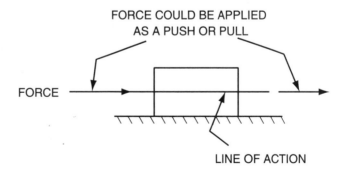

Figure A.1   Horizontal force acting on a body at rest

The force is shown on the line on which it acts (the **line of action**) and the arrowhead shows the direction. The effect on the object is the same whether it is pushed or pulled, so it does not matter where the force is shown on the line of action as long as the direction is indicated.

The force is represented graphically by an arrow pointing in the direction of the force. The length of the arrow represents the magnitude of the force. The vector is drawn to some scale (e.g., 1 cm = 10 N, 1 cm on the paper represents a force of 10 N). The magnitude is read directly from the scale so there is no arithmetic to do. More on this later. That is all there is to it. Draw a line of specific length in the direction of the force.

Now let's look at adding vectors.

Adding Vectors

We usually deal with more than one vector so we must know how to combine them. When two forces act on an object we may want to replace these forces with one equivalent force by adding them. Vectors are added by joining them "nose-to-tail." Figure A.2 shows two forces, A and B, acting at a point. The forces are on their lines of action but are not drawn to scale.

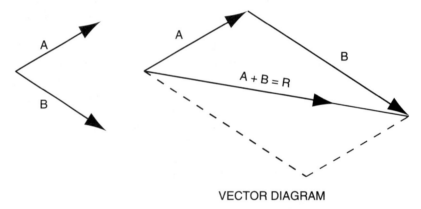

VECTOR DIAGRAM

Figure A.2   Adding vectors

Vector A representing force A, is drawn to scale, at the same angle as the force. The length represents the magnitude of the force. The vector B is drawn at the same angle as force B, starting from the end of A. The resultant is found by joining the starting point to the end point. Forces A and B could be replaced by one force of magnitude R, acting in the direction of the resultant (the arrow is shown in mid-line for clarity). Vectors can be added in any order. The result is the same if force B is drawn first and A is added at the end, as shown in Figure A.2. This is called the **parallelogram law**.

Now let's look at a practical application of this information.

Example A1.1

A boat leaves a pier on one side of a river flowing at 0.5 m/s in the direction shown. The operator heads straight for the opposite shore. If the river is 500 m wide where will the boat hit the opposite shore if it moves at 1 m/s ? Figure A.3 shows the layout.

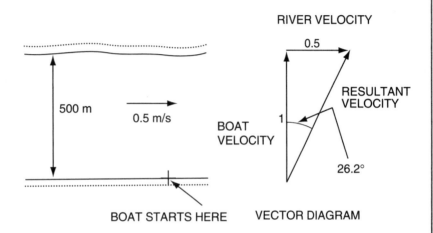

Figure A.3

The boat heads toward the opposite shore at a velocity of 1 m/s (the direction and magnitude are known), but the river will carry it downstream. The resultant velocity is the sum of boat velocity and river velocity. Figure A.3 shows the addition of these velocities and the resulting boat velocity.

The boat moves downstream as it crosses the river and hits the opposite bank at a point 246 m (500 tan 26.2°) downstream of the starting point.

The process of adding vectors is the same for any number of vectors. Figure A.4 shows a frame loaded with four forces. This is a **space diagram**, that shows the frame is drawn to scale (e.g., 1:100) and the forces are shown on their lines of action. The forces are not drawn to scale. The space diagram shows where they act and the direction. The space scale must be shown.

The resultant of these forces is found by adding the force vectors nose-to-tail in a vector diagram. Each force is drawn to scale (e.g., 1 cm = 2 kN) in the same direction as the line of action. The forces can be added in any order.

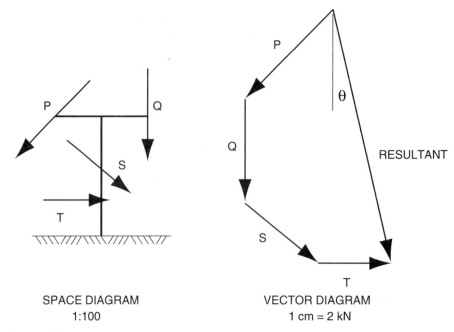

SPACE DIAGRAM
1:100

VECTOR DIAGRAM
1 cm = 2 kN

Figure A.4

Measure the resultant using the appropriate scale to find the magnitude. The direction, in this case, is downward at angle ϕ from the vertical. Now let's move on to look at freebody diagrams.

Freebody Diagrams

The first step in any analysis involving forces is to identify the forces and their lines of action. A **freebody diagram** is used to do this. A freebody diagram shows an object as if it were floating in air, supported only by the forces acting on it. Figure A.5 shows a beam supported at each end, and carrying a load, W. The left end is supported by a pin joint, and the right end by a roller support. A pin joint can take a force in any direction, a roller support can carry a load only in a normal direction (perpendicular to the surface). The weight of the beam is small relative to the applied force and can be neglected.

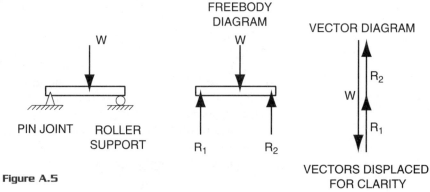

Figure A.5

A freebody diagram is drawn by removing the supports and replacing them with forces. The other force acting is the load, W. As a check, and to plan the next step, sketch a vector diagram. Since the object is in equilibrium (not moving) the forces acting on it must balance and the vector diagram must close. (It must start and finish at the same point and there is no resultant force.) If it does not close, you know there is an error somewhere. The three forces are all vertical so they have been displaced for clarity in Figure A.5. At this stage, a sketch is sufficient to determine whether the diagram closes.

Now change the loading and see how the fact that the system is in equilibrium can be used. Figure A.6 shows the load and freebody diagrams for a beam with a non-vertical load, L, added.

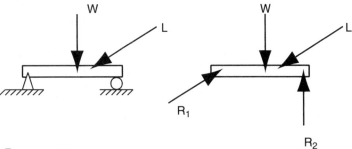

Figure A.6

The direction of the force at the left end is not known, but we know it is not vertical, since there must be some force to balance the non-vertical load, L. If this force is not balanced the system would move to the left, and, since it is not moving, there must be some force acting to the right.

The direction of force R_1 can be found by drawing the vector diagram. The directions of the other three forces are known. We can sketch the vector diagram to see how R_1 can be found. Since we have no values all we can do is a sketch, but this will be sufficient to determine what to do. Figure A.7 shows how the vector diagram is constructed.

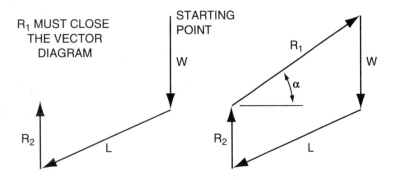

Figure A.7

Forces W, L, and R_2 can be added because their directions are known. Since the system is in equilibrium, the vector diagram must close. Therefore, the line representing force R_1 must go from the nose of R_2 back to the starting point. This would establish the magnitude and direction of R_1, if the magnitudes of the other forces were known. Sketching the vector diagram shows how to approach the problem. The vector diagram, drawn to scale, is used to find the magnitude and direction of R_1.

Now let's look at friction.

Friction

If a body is pushed on a surface, there is a friction force opposing the motion. If the force pushing the object is removed, the friction force will bring the object to a stop. Friction always opposes motion. Friction force depends only on the **coefficient of static friction**, μ, and the weight of the object. The areas of the two surfaces have no effect on friction force. The coefficient of friction depends on the two surfaces in contact. We will deal with the coefficient of static friction that applies when a body is just about to move. A **coefficient of sliding friction** applies when an object is sliding on a surface, and a **coefficient of rolling friction** applies to a wheel rolling on a surface. Some typical coefficients of static friction are:

Hard steel on hard steel	0.78
Mild steel on mild steel	0.74
Aluminum on mild steel	0.61
Brass on mild steel	0.51
Teflon on steel	0.04

[Source: Baumeister, T. and L.S. Marks (eds.). *Standard Handbook for Mechanical Engineers*, Seventh Edition. New York: McGraw-Hill Book Company, 1967.]

Consider a block on a flat surface. The coefficient of friction for the two surfaces is μ. If a small horizontal force is applied to the block, it will not move because the force is less than the friction force opposing motion. If the force is gradually increased, the block will move when the horizontal force is large enough to just overcome the friction force. The limiting friction force depends on the coefficient of friction and the normal force of the block on the surface. (The normal force may, or may not, be equal to the weight of the block.) When motion is impending (the block is just about to move) friction force, f, is equal to:

$$f = \mu N,$$

where N is the normal force. Figure A.8 shows the block, freebody, and vector diagrams. The line of action of the friction force is on the surface.

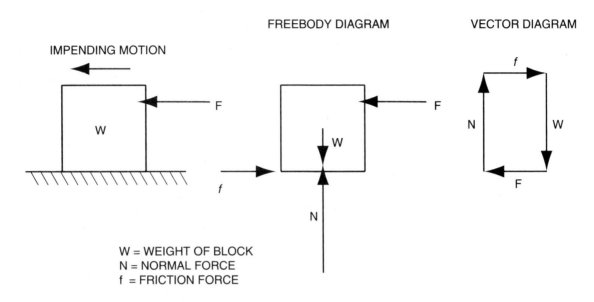

FREEBODY DIAGRAM   VECTOR DIAGRAM

IMPENDING MOTION

W = WEIGHT OF BLOCK
N = NORMAL FORCE
f = FRICTION FORCE

Figure A.8

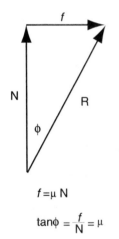

$f = \mu\, N$

$\tan\phi = \dfrac{f}{N} = \mu$

Figure A.9

Normal and friction force can be added to give a resultant and if the coefficient of friction is known, the direction of the resultant can easily be found. Figure A.9 shows how this is done.

The fact that the direction can be found, even though the magnitude of the resultant is not known, is useful in problems involving friction. The actual value of ϕ, and the magnitude of the resultant are not usually important and are seldom required. Since the tangent is known ($\tan \phi = \mu$), the angle can be drawn, giving the direction of the resultant. Figure A.10 shows how to construct an angle when you know its tangent.

$$\tan \phi = \frac{f}{N} = \mu$$

TO CONSTRUCT AN ANGLE WITH TANGENT 0.27

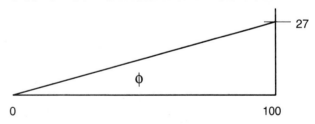

Figure A.10   Constructing an angle knowing the tangent

1. Draw a line 100 units long. Any scale can be used, but a long line (about 100 mm) gives better accuracy.
2. Erect a perpendicular at the end of the line.
3. Measure 27 units (at the same scale) on the perpendicular and join this point with the starting point.

It is more accurate to construct the angle this way than calculate and measure it with a protractor. The angle corresponding to $\mu = 0.27$ is 15.11°, which is difficult to read accurately on a protractor. One hundred, and twenty-seven are easy to read on a scale.

Let's see how we can use all this to draw a vector diagram knowing the magnitude of one force and all of the directions.

Example A1.2

Figure A.11 shows two blocks connected by a rigid link. Determine the weight of block A that will just prevent slipping in the position shown. Block B weighs 220 N and the coefficient of friction between all surfaces is 0.22.

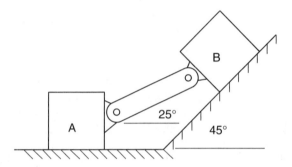

Figure A.11

We want to find the weight of block A that will just hold the system in the position shown. If the weight of block A is reduced a small amount, the system will move to the left. The following are the steps to find the weight.

1. Sketch the freebody diagram for all components, blocks A and B, and the link, to identify all forces acting and their directions for the condition of motion impending to the left (see Figure A.12).

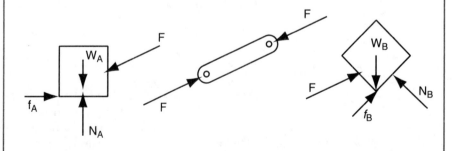

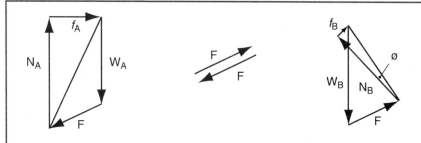

Figure A.12   Freebody and vector diagrams

2. Sketch vector diagrams for all components to see how scale drawings should be constructed to find the required information. Since these are sketches, there is no scale and no values are shown. Show the angular relation (ϕ) between the friction force and the associated normal force (see Figure A.12).

    While vectors can be added in any order, some combinations do not help in solving the problem. Always add the friction force to the end of the normal force so the angle ϕ can be shown.

3. If the vector diagram for A is drawn to scale, W_A can be found, but there is not enough information to do this. The directions of all forces are known, but no magnitudes. If the magnitude of one force were known the vector diagram could be drawn, and W_A found by measuring the length of the line representing W_A.

    The force on the link, F, is common to vector diagrams for both A and B, and can be found using the vector diagram for block B.

4. Draw the vector diagram for B. Since only W_B is known, it must be the starting point (see Figure A.13). A scale of 1 cm = 50 N is used. Values can be read directly off the 1:5 metric scale.

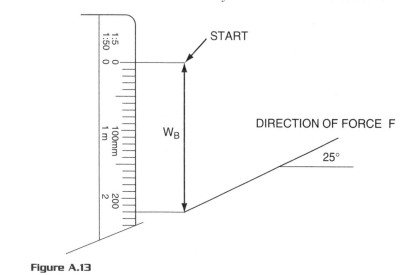

Figure A.13

a) Draw W_B to scale.

b) Add force F at the end of W_B. The starting point for F is known, but not the length. Draw it any length in the correct direction.

c) The vector diagram must close, since the system is in equilibrium. We must get from some point on F back to the starting point by adding friction force, f_B, and normal force, N_B. The directions of these are known, but where they start on F is not. Work back from the starting point of the vector diagram (the tail of W_B). Look at the sketch to see how this is done. The sketch is repeated in Figure A.14.

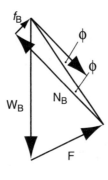
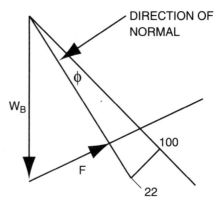

Figure A.14

d) Construct the angle ϕ starting at the end of W_B by drawing a line, 100 units long, parallel to the normal. At the end of this line, draw a perpendicular 22 units long (μ = 0.22). Complete the triangle to find where the resultant intersects F. The magnitude of F is now known. There is no need to find the magnitude of N_B or f_B. There is no need to find the magnitude of F either, since all we need is the scale length.

e) Draw the vector diagram for A. Figure A.15 shows the process. Draw a vertical line representing the direction of W_A. The starting point is not known. Add F at the end of W_A (see Figure A.15a). At the nose of F, draw a line in the direction of the resultant of f_A and N_A using the method described above. The intersection of the resultant and W_A gives the length of W_A. Measure the length to scale. W_A = 440 N (see Figure A.15b).

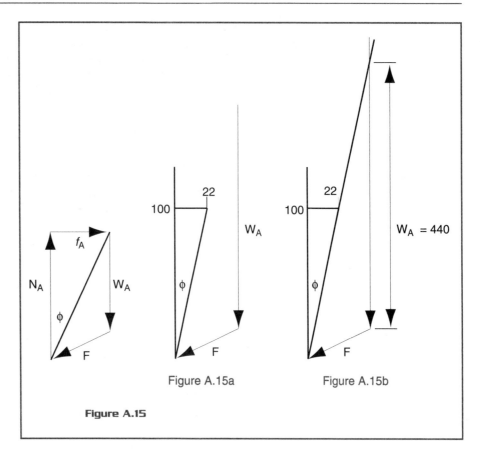

Figure A.15a   Figure A.15b

Figure A.15

Now let's look at finding the line of action of the resultant.

Finding the Line of Action of the Resultant

It is usually necessary to determine what effect forces have on a system. To do this, all lines of action must be known. For example, the net force acting on a system may result in it moving in an undesirable direction, or in failure. It is often necessary to find where the line of action of the resultant is or where some force acts.

We saw how two vectors can be added to find a resultant. The resultant acts through the intersection of the lines of action of the vectors. Similarly a vector quantity can also be represented by two components. You may have used this technique to replace a vector quantity with Cartesian components (components aligned with x and y axes) as shown in Figure A.16.

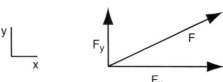

Figure A.16   Cartesian components

Cartesian components are a special case of components at 90°, but components can be at any angle. Figure A.17 shows a force, F, replaced by arbitrary components. The line on top of the letters in the equation indicates that this is a vector addition. The arrowhead on the resultant has been moved away from the end of the line for clarity.

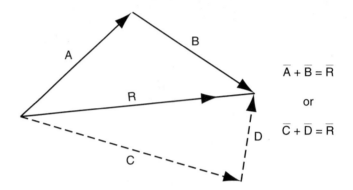

$$\overline{A} + \overline{B} = \overline{R}$$

or

$$\overline{C} + \overline{D} = \overline{R}$$

Figure A.17

The use of vector components to locate the line of action of a resultant can be shown best by an example.

Example A1.3

A barge, shown in Figure A.18, is acted on by three forces. These forces move the barge. How it moves depends on the location of the line of action of the resultant. The craft could move in the direction of the resultant or it could rotate, depending on the location of the line of action.

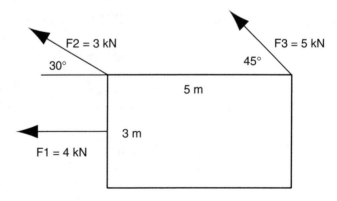

Figure A.18

1. Find the magnitude and the direction of the resultant by adding the forces. Figure A.19 shows the vector diagram used to find this information.

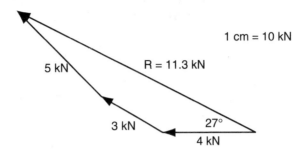

Figure A.19

2. Find two components of resultant, R. The line of action of R passes through the intersection of these components. Figure A.20 shows the components of R. One component is F_3. The other component, R_1, is the resultant of F_1 and F_2.

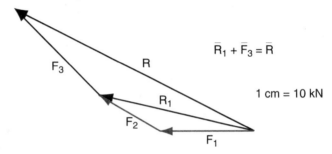

Figure A.20

3. Replace F_1 and F_2 by resultant, R_1, on the space diagram. The resultant acts through the intersection of the lines of action of F_1 and F_2 (see Figure A.21).

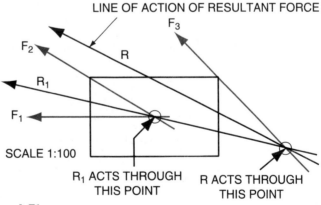

Figure A.21

4. Find the intersection of R_1 and F_3 by projecting R_1 and F_3. Resultant R passes through this point (see Figure A.21).

5. Draw R on the space diagram.

Example A1.4

Sketch freebody and vector diagrams for the three components of the cargo sling.

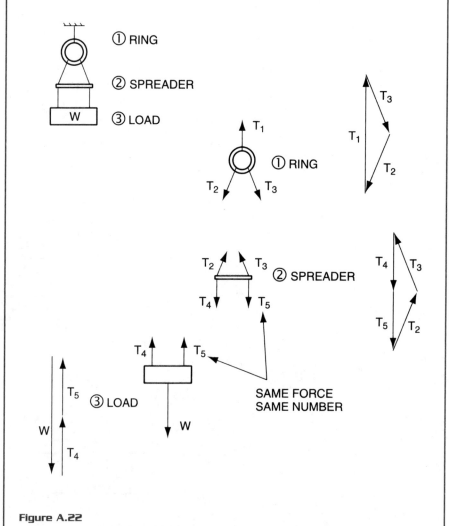

CARGO SLING

① RING

② SPREADER

③ LOAD

Figure A.22

Problems

For all problems, sketch the freebody and vector diagrams for all components. Then draw vector diagrams, to scale, to find the required information.

1. Sketch freebody and vector diagrams for the objects shown in Figure A.23. Sketches should be done for all numbered components and not for the assembly, as shown in Example A1.4. Ignore the weight of the part unless weight is indicated by W.

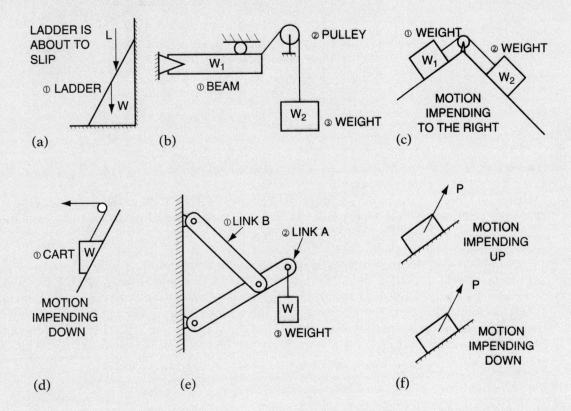

Figure A.23

2. Figure A.24 shows a system of wedges, A and B, used for positioning heavy parts. The part weighs 8.5 kN and the coefficient of friction between the part and the floor is 0.18. The weight of A and B is negligible and the coefficient of friction between A and B and any surface over which they slide is 0.12. Find the force, P, that will just move the part to the right. The scale is 1 cm = 0.5 kN.

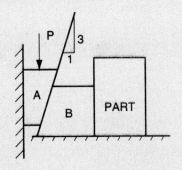

Figure A.24

3. Figure A.25 shows a system of wedges intended to move C to the right against a force of 4 kN. The weight of parts A, B, and C is negligible. The coefficient of friction for all surfaces is 0.18. Find the force, P, that will just cause the system to move. The scale is 1 cm = 200 N.

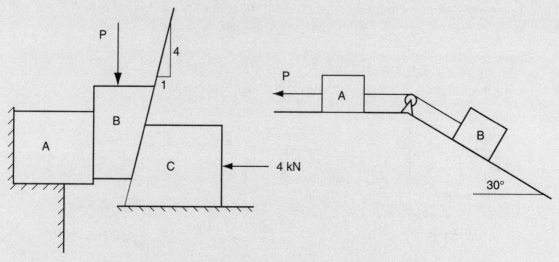

Figure A.25   **Figure A.26**

4. What is the force, P, that will just move the system in Figure A.26 to the left? Blocks A and B weigh 1000 N and 1500 N respectively. The coefficient of friction for all surfaces is 0.23. Ignore friction in the pulley. The scale is 1 cm = 200 N.

5. A mechanical system can be represented as shown in Figure A.27. A force, P, that will just move the system to the left must be applied to A. Blocks A and B weigh 700 N and 950 N respectively. The coefficient of friction is 0.2 for all surfaces. The pulley is frictionless and the rope connecting the blocks is parallel to the surface. Specify the magnitude and direction of the minimum force, P, that will just move the system to the left. The broken line indicates that the line of action is not known and must be specified. The scale is 1 cm = 100 N.

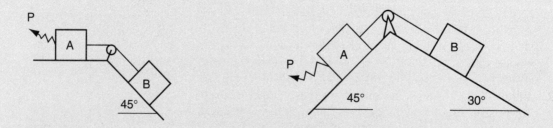

Figure A.27   **Figure A.28**

6. Find the minimum force, P, that will just move the system shown in Figure A.28 to the left. Block A weighs 450 N and block B weighs 1400 N. The coefficient of friction for all surfaces is 0.18. The broken line showing where force P is applied indicates that the line of action is not known and must be determined. The scale is 1 cm = 100 N.

7. What is the weight of A that will just cause the system described in problem 6 to slide to the left with no applied force?

8. For the system described in problem 6, find the minimum force, P, that will just prevent the system slipping to the right. The scale is 1 cm = 100 N.

9. Figure A.29 shows a proposed mechanism for raising a table and a part to be machined. The force, F, will be applied by a hydraulic cylinder. Force, F, must be known so that the cylinder can be specified. The table and the part to be machined weigh 500 N. The coefficient of friction for all surfaces is 0.25. The weights of parts A and B are negligible. The scale is 1 cm = 50 N.
    Find the force that will just raise the table and part.
    Is this proposed mechanism a good one for this application?

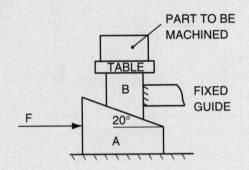

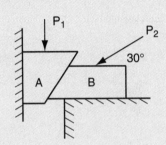

Figure A.29   **Figure A.30**

10. Block A weighs 200 N and B weighs 300 N (see Figure A.30). The coefficient of friction for all surfaces is 0.23. Force P_2 is 550 N and P_1 is just sufficient to move A down. Find the force P_1 that will just cause A to move. The scale is 1 cm = 100 N.

11. A worker claims he was injured when climbing up the ladder shown in Figure A.31 because it slipped. Investigate this case and determine whether the worker has a valid claim. The worker weighs 800 N. The coefficient of friction between the ladder and the ground is 0.62 and between the ladder and the wall, 0.38. The ladder weighs 100 N. The space scale is 1:50. The force scale is 1 cm = 100 N.

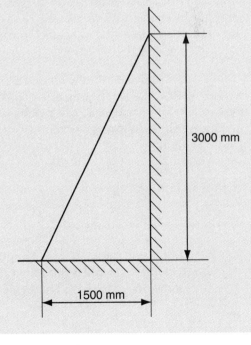

Figure A.31

12. A building mounted on skids is to be moved across a smooth, level parking lot (see Figure A.32). Three tractors will be used to move the building. One tractor applies a force of 9 kN, due west at corner A. Another applies 20 kN due west at corner B. A third tractor applies a force of 13 kN northwest at C. The combined effort is sufficient to move the building slowly across the parking lot.

   Determine the magnitude and direction of the resultant of the three forces. Dimension the distance from C where the line of action of the resultant crosses CD. The space scale is 1:200. The force scale is 1 cm = 2 kN.

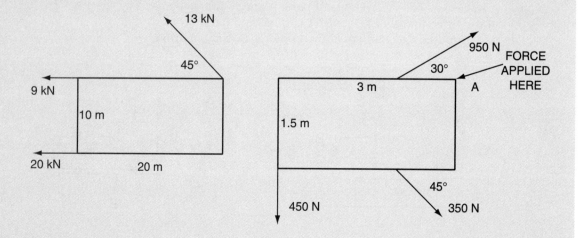

Figure A.32   **Figure A.33**

13. A raft is acted on by the forces shown in Figure A.33. The raft must be moved alongside a pier by the minimum possible force applied at A. The resultant of all forces acting on the raft must be to the right and downward at 30°.
    a) Find the magnitude and direction of the force applied at A.
    b) Determine the magnitude of the resultant force.
    c) Show where the line of action acts on the space diagram.

14. Figure A.34 shows a 3 kN cylinder being pulled up a ramp. The coefficient of friction at A and B is 0.4. The ramp weighs 1.2 kN. At some point in the travel, the ramp will slip. How far up the ramp can the cylinder be pulled before the ramp slips? Ignore friction between the cylinder and the ramp.
    a) Sketch the freebody diagrams for the cylinder and the ramp.
    b) Sketch the vector diagrams for the cylinder and the ramp.
    c) Draw the vector diagram for the cylinder and determine the normal force between the cylinder and ramp. The scale is 1 cm = 500 N.
    d) Draw the vector diagram for the ramp and determine RA and RB. The scale is 1 cm = 500 N.

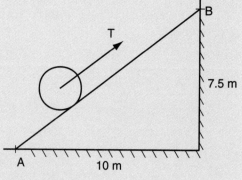

Figure A.34

e) How far from the bottom of the ramp is the center of the cylinder when the ramp slips? Show this distance along the ramp.

15. A load-limiting device is shown in Figure A.35. It is designed to prevent the load, F, from exceeding 50 kN. If this load is exceeded, the system will slip. Specify the coefficient of friction required to meet these conditions. The coefficient of friction is the same for all sliding surfaces. The scale is 1 cm = 10 kN.

W_A = 70 kN
W_B = 60 kN
α = 30°

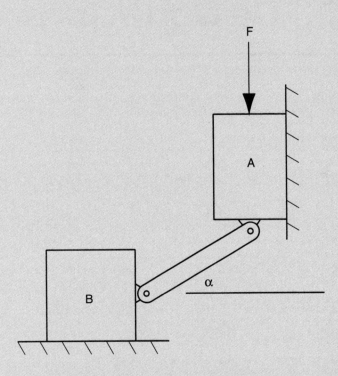

Figure A.35

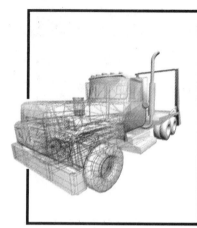

Appendix B

Common Sectioning Symbols and Designations

There are sectioning symbols for many materials. The following are some of the common ones.

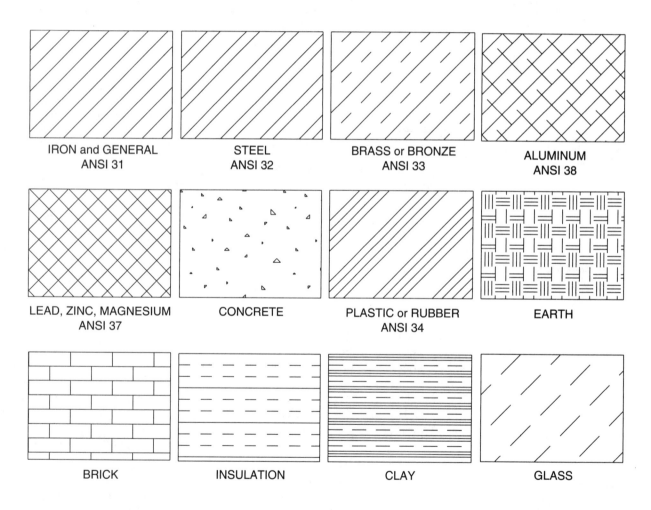

IRON and GENERAL
ANSI 31

STEEL
ANSI 32

BRASS or BRONZE
ANSI 33

ALUMINUM
ANSI 38

LEAD, ZINC, MAGNESIUM
ANSI 37

CONCRETE

PLASTIC or RUBBER
ANSI 34

EARTH

BRICK

INSULATION

CLAY

GLASS

Appendix C
Standards List

There are hundreds of standards in engineering. This is a very small list, pertaining mainly to drawings and fasteners. More information on standards can be found from the following web sites. Standards can be ordered over the Internet from most of these organizations. There are many organizations that issue standards, but only a few are listed here.

American National Standards Institute	www.ansi.org
American Society for Testing and Materials	www.astm.org
Canadian Standards Association	www.csa-international.org
Institute of Electrical and Electronics Engineers	www.standards.ieee.org
International Standards Organization	www.iso.ch
National Standards System Network	www.nssn.org
Society of Automotive Engineers	www.sae.org

ASME Y1.1-1989
Abbreviations for Use on Drawings and in Text
Standard abbreviations for use on engineering drawings

ASME Y14.1-1995
Decimal Inch Sheet Size and Format
Sheet size and format for engineering drawings

ASME Y14.1M-1995
Metric Drawing Sheet Size and Format
Metric sheet size and format for engineering drawings

ASME Y14.2M-1992(R1998)
Line Conventions and Lettering
Line and lettering practices used on engineering drawings

ASME Y14.3M-1994
Multiview and Sectional View Drawings
Requirements for selection and arrangement of orthographic views, section views, auxiliary views, and conventional drawing practices

ASME Y14.4M-1989(R1994)
Pictorial Drawing
Defines and illustrates the use of various kinds of pictorial drawings used in mechanical drawings

ASME Y14.5M-1994
Dimensioning and Tolerancing
Methods of dimensioning and specifying tolerances on engineering drawings

CSA B78.1
Technical Drawing, General Principles

CSA B78.2
Dimensioning and Tolerancing of Technical Drawings

ASME Y14.6-1978(R1993)
Screw Thread Representation
Standards for representing, dimensioning, and specifying screw threads

ASME Y14.6M-1981(R1998)
Screw Thread Representation (Metric Supplement)
Standards for representing, dimensioning, and specifying metric threads

ASME Y32.2.3-1949(R1994)
Graphic Symbols for Pipe Fittings, Valves, and Piping

ASME Y32.2.4-1949(R1998)
Graphic Symbols for Heating, Ventilating, and Air Conditioning

ASME Y32.2.6-1950(R1998)
Graphic Symbols for Heat-Power Apparatus

ASME Y32.4-1977(R1994)
Graphic Symbols for Plumbing Fixtures for Diagrams Used in Architecture and Building Construction

ASME Y32.7-1972(R1994)
Graphic Symbols for Railway Maps and Profiles

ASME Y32.10-1967(R1994)
Graphic Symbols for Fluid Power Diagrams

ASME Y32.11-1961(R1998)
Graphic Symbols for Process Diagrams in Petroleum and Chemical Industries

IEEE315A-86
Graphic Symbols for Electrical and Electronic Diagrams

IEEE91-84
Graphic Symbols for Logic Diagrams

AISC S340 (American Institute of Steel Construction)
Metric Properties of Structural Steel Shapes with Dimensions According to ASTM A6M

AISC P648
Fundamentals of (Metric) Structural Shop Drafting—CISC (Revised December 1998)

ASME B1.1-1998
Unified Inch Screw Threads (UN and UNR Thread Form)
Designations and other information for unified threads

ASME B1.13-1995
Metric Screw Threads—M Profile
General standards for metric threads

ISO 262:1973
ISO General Purpose Metric Screw Threads—Selected Sizes for Screws, Bolts, and Nuts

ISO 68:1973
ISO General Purpose Screw Threads—Basic Profile

ISO 724:1993
ISO General Purpose Screw Threads—Basic Dimensions

ASME B18.2.1-1996
Square and Hex Bolts and Screws (Inch Series)
General information and dimensions

ASME B18.2.2-1987(R1993)
Square and Hex Nuts
General information and dimensions

ASME B18.3-1998
Socket, Cap, Shoulder, and Set Screws, Hex, and Spline Keys
General information and dimensions

ASME B18.5-1990(R1998)
Round Head Bolts (Inch Series)
General information and dimensions

ASME B18.6.2-1972(R1993)
Slotted Head Cap Screws, Square Head Set Screws, and Slotted Headless Set Screws
General information and dimensions

ASME B18.6.3-1972(R1997)
Machine Screws and Machine Screw Nuts
General information and dimensions

ASME B1.20.1-1983(R1992)
Pipe Threads, General Purpose (Inch)
Dimensions and gauging of pipe threads

ASME-A13.1-81R85
Scheme for the Identification of Piping Systems

Appendix D

Standard Dimensions (Cap and Machine Screws)

ISO Metric Screw Thread Standard Series

Nominal Size Dia. (mm) Column[a]			Pitches (mm) Series with Graded Pitches		Series with Constant Pitches												Nominal Size Dia. (mm)
1	2	3	Coarse	Fine	6	4	3	2	1.5	1.25	1	0.75	0.5	0.35	0.25	0.2	
0.25			0.075	—	—	—	—	—	—	—	—	—	—	—	—	—	0.25
0.3			0.08	—	—	—	—	—	—	—	—	—	—	—	—	—	0.3
	0.35		0.09	—	—	—	—	—	—	—	—	—	—	—	—	—	0.35
0.4			0.1	—	—	—	—	—	—	—	—	—	—	—	—	—	0.4
	0.45		0.1	—	—	—	—	—	—	—	—	—	—	—	—	—	0.45
0.5			0.125	—	—	—	—	—	—	—	—	—	—	—	—	—	0.5
	0.55		0.125	—	—	—	—	—	—	—	—	—	—	—	—	—	0.55
0.6			0.15	—	—	—	—	—	—	—	—	—	—	—	—	—	0.6
	0.7		0.175	—	—	—	—	—	—	—	—	—	—	—	—	—	0.7
0.8			0.2	—	—	—	—	—	—	—	—	—	—	—	—	—	0.8
	0.9		0.225	—	—	—	—	—	—	—	—	—	—	—	—	—	0.9
			0.25	—	—	—	—	—	—	—	—	—	—	—	—	0.2	1
	1.1		0.25	—	—	—	—	—	—	—	—	—	—	—	—	0.2	1.1
1.2			0.25	—	—	—	—	—	—	—	—	—	—	—	—	0.2	1.2
	1.4		0.3	—	—	—	—	—	—	—	—	—	—	—	—	0.2	1.4
1.6			0.35	—	—	—	—	—	—	—	—	—	—	—	—	0.2	1.6
	1.8		0.35	—	—	—	—	—	—	—	—	—	—	—	—	0.2	1.8
2			0.4	—	—	—	—	—	—	—	—	—	—	—	0.25	—	2
	2.2		0.45	—	—	—	—	—	—	—	—	—	—	—	0.25	—	2.2
2.5			0.45	—	—	—	—	—	—	—	—	—	—	0.35	—	—	2.5
3			0.5	—	—	—	—	—	—	—	—	—	—	0.35	—	—	3
	3.5		0.6	—	—	—	—	—	—	—	—	—	—	0.35	—	—	3.5
4			0.7	—	—	—	—	—	—	—	—	—	0.5	—	—	—	4
	4.5		0.75	—	—	—	—	—	—	—	—	—	0.5	—	—	—	4.5
5			0.8	—	—	—	—	—	—	—	—	—	0.5	—	—	—	5
		5.5	—	—	—	—	—	—	—	—	—	—	0.5	—	—	—	5.5
6			1	—	—	—	—	—	—	—	—	0.75	—	—	—	—	6
		7	1	—	—	—	—	—	—	—	—	0.75	—	—	—	—	7
8			1.25	1	—	—	—	—	—	—	1	0.75	—	—	—	—	8
		9	1.25	—	—	—	—	—	—	—	1	0.75	—	—	—	—	9
10			1.5	1.25	—	—	—	—	—	1.25	1	0.75	—	—	—	—	10
		11	1.5	—	—	—	—	—	—	—	1	0.75	—	—	—	—	11
12			1.75	1.25	—	—	—	—	1.5	1.25	1	—	—	—	—	—	12
	14		2	1.5	—	—	—	—	1.5	1.25[b]	1	—	—	—	—	—	14
		15	—	—	—	—	—	—	1.5	—	1	—	—	—	—	—	15
16			2	1.5	—	—	—	—	1.5	—	1	—	—	—	—	—	16
		17	—	—	—	—	—	—	1.5	—	1	—	—	—	—	—	17
	18		2.5	1.5	—	—	—	2	1.5	—	1	—	—	—	—	—	18
20			2.5	1.5	—	—	—	2	1.5	—	1	—	—	—	—	—	20
	22		2.5	1.5	—	—	—	2	1.5	—	1	—	—	—	—	—	22

[a] Thread diameter should be selected from columns 1, 2 or 3, with preference being in that order.
[b] Pitch 1.25 mm in combination with diameter 14 mm has been included for sparkplug applications.
[c] Diameter 35 mm has been included for bearing locknut applications.

The use of pitches shown in parentheses should be avoided wherever possible.

The pitches enclosed in the bold frame, together with the corresponding nominal diameters in columns 1 and 2, are those combinations which have been established by ISO Recommendations as a selected "coarse" and "fine" series for commercial fasteners.

ISO Metric Screw Thread Standard Series (cont.)

Nominal Size Dia. (mm)			Pitches (mm)														Nominal Size Dia. (mm)
Column[a]			Series with Graded Pitches		Series with Constant Pitches												
1	2	3	Coarse	Fine	6	4	3	2	1.5	1.25	1	0.75	0.5	0.35	0.25	0.2	
24			3	2	—	—	—	2	1.5	—	1	—	—	—	—	—	24
		25	—	—	—	—	—	2	1.5	—	1	—	—	—	—	—	25
		26	—	—	—	—	—	—	1.5	—	1	—	—	—	—	—	26
	27		3	2	—	—	—	2	1.5	—	1	—	—	—	—	—	27
		28	—	—	—	—	—	2	1.5	—	1	—	—	—	—	—	28
30			3.5	2	—	—	(3)	2	1.5	—	1	—	—	—	—	—	30
		32	—	—	—	—	—	2	1.5	—	—	—	—	—	—	—	32
	33		3.5	2	—	—	(3)	2	1.5	—	—	—	—	—	—	—	33
		35[c]	—	—	—	—	—	—	1.5	—	—	—	—	—	—	—	35[c]
36			4	3	—	—	—	2	1.5	—	—	—	—	—	—	—	36
		38	—	—	—	—	—	—	1.5	—	—	—	—	—	—	—	38
	39		4	3	—	—	—	2	1.5	—	—	—	—	—	—	—	39
		40	—	—	—	—	3	2	1.5	—	—	—	—	—	—	—	40
42			4.5	3	—	4	3	2	1.5	—	—	—	—	—	—	—	42
	45		4.5	3	—	4	3	2	1.5	—	—	—	—	—	—	—	45
48			5	3	—	4	3	2	1.5	—	—	—	—	—	—	—	48
		50	—	—	—	—	3	2	1.5	—	—	—	—	—	—	—	50
	52		5	3	—	4	3	2	1.5	—	—	—	—	—	—	—	52
		55	—	—	—	4	3	2	1.5	—	—	—	—	—	—	—	55
56			5.5	4	—	4	3	2	1.5	—	—	—	—	—	—	—	56
		58	—	—	—	4	3	2	1.5	—	—	—	—	—	—	—	58
	60		5.5	4	—	4	3	2	1.5	—	—	—	—	—	—	—	60
		62	—	—	—	4	3	2	1.5	—	—	—	—	—	—	—	62
64			6	4	—	4	3	2	1.5	—	—	—	—	—	—	—	64
		65	—	—	—	4	3	2	1.5	—	—	—	—	—	—	—	65
	68		6	4	—	4	3	2	1.5	—	—	—	—	—	—	—	68
		70	—	—	6	4	3	2	1.5	—	—	—	—	—	—	—	70
72			—	—	6	4	3	2	1.5	—	—	—	—	—	—	—	72
		75	—	—	—	4	3	2	1.5	—	—	—	—	—	—	—	75
	76		—	—	6	4	3	2	1.5	—	—	—	—	—	—	—	76
		78	—	—	—	—	—	2	—	—	—	—	—	—	—	—	78
80			—	—	6	4	3	2	1.5	—	—	—	—	—	—	—	80
		82	—	—	—	—	—	2	—	—	—	—	—	—	—	—	82
	85		—	—	6	4	3	2	—	—	—	—	—	—	—	—	85
90			—	—	6	4	3	2	—	—	—	—	—	—	—	—	90

Cont.

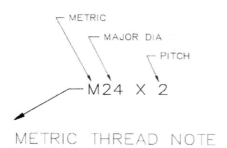

METRIC THREAD NOTE

ISO Metric Screw Thread Standard Series (cont.)

Nominal Size Dia. (mm) Column[a]			Series with Graded Pitches		Series with Constant Pitches												Nominal Size Dia. (mm)
1	2	3	Coarse	Fine	6	4	3	2	1.5	1.25	1	0.75	0.5	0.35	0.25	0.2	
	95		—	—	6	4	3	2	—	—	—	—	—	—	—	—	95
100			—	—	6	4	3	2	—	—	—	—	—	—	—	—	100
	105		—	—	6	4	3	2	—	—	—	—	—	—	—	—	105
110			—	—	6	4	3	2	—	—	—	—	—	—	—	—	110
	115		—	—	6	4	3	2	—	—	—	—	—	—	—	—	115
	120		—	—	6	4	3	2	—	—	—	—	—	—	—	—	120
125			—	—	6	4	3	2	—	—	—	—	—	—	—	—	125
	130		—	—	6	4	3	2	—	—	—	—	—	—	—	—	130
		135	—	—	6	4	3	2	—	—	—	—	—	—	—	—	135
140			—	—	6	4	3	2	—	—	—	—	—	—	—	—	140
		145	—	—	6	4	3	2	—	—	—	—	—	—	—	—	145
	150		—	—	6	4	3	2	—	—	—	—	—	—	—	—	150
		155	—	—	6	4	3	—	—	—	—	—	—	—	—	—	155
160			—	—	6	4	3	—	—	—	—	—	—	—	—	—	160
		165	—	—	6	4	3	—	—	—	—	—	—	—	—	—	165
	170		—	—	6	4	3	—	—	—	—	—	—	—	—	—	170
		175	—	—	6	4	3	—	—	—	—	—	—	—	—	—	175
180			—	—	6	4	3	—	—	—	—	—	—	—	—	—	180
		185	—	—	6	4	3	—	—	—	—	—	—	—	—	—	185
	190		—	—	6	4	3	—	—	—	—	—	—	—	—	—	190
		195	—	—	6	4	3	—	—	—	—	—	—	—	—	—	195
200			—	—	6	4	3	—	—	—	—	—	—	—	—	—	200
		205	—	—	6	4	3	—	—	—	—	—	—	—	—	—	205
	210		—	—	6	4	3	—	—	—	—	—	—	—	—	—	210
220			—	—	6	4	3	—	—	—	—	—	—	—	—	—	220
		225	—	—	6	4	3	—	—	—	—	—	—	—	—	—	225
		230	—	—	6	4	3	—	—	—	—	—	—	—	—	—	230
		235	—	—	6	4	3	—	—	—	—	—	—	—	—	—	235
	240		—	—	6	4	2	—	—	—	—	—	—	—	—	—	240
		245	—	—	6	4	3	—	—	—	—	—	—	—	—	—	245
250			—	—	6	4	3	—	—	—	—	—	—	—	—	—	250
		255	—	—	6	4	—	—	—	—	—	—	—	—	—	—	255
	260		—	—	6	4	—	—	—	—	—	—	—	—	—	—	260
		265	—	—	6	4	—	—	—	—	—	—	—	—	—	—	265
		270	—	—	6	4	—	—	—	—	—	—	—	—	—	—	270
		275	—	—	6	4	—	—	—	—	—	—	—	—	—	—	275
280			—	—	6	4	—	—	—	—	—	—	—	—	—	—	280
		285	—	—	6	4	—	—	—	—	—	—	—	—	—	—	285
		290	—	—	6	4	—	—	—	—	—	—	—	—	—	—	290
		295	—	—	6	4	—	—	—	—	—	—	—	—	—	—	295
	300		—	—	6	4	—	—	—	—	—	—	—	—	—	—	300

[a] Thread diameter should be selected from columns 1, 2, or 3; with preference being in that order.

American Standard Square Bolts and Nuts

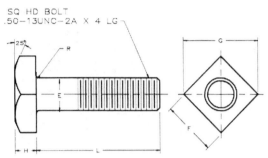

SQ HD BOLT
.50−13UNC−2A X 4 LG

Dimensions of Square Bolts

Nominal Size or Basic Product Dia		Body Dia E	Width Across Flats F			Width Across Corners G		Height H			Radius of Fillet R
		Max	Basic	Max	Min	Max	Min	Basic	Max	Min	Max
1/4	0.2500	0.260	3/8	0.3750	0.362	0.530	0.498	11/64	0.188	0.156	0.031
5/16	0.3125	0.324	1/2	0.5000	0.484	0.707	0.665	13/64	0.220	0.186	0.031
3/8	0.3750	0.388	9/16	0.5625	0.544	0.795	0.747	1/4	0.268	0.232	0.031
7/16	0.4375	0.452	5/8	0.6250	0.603	0.884	0.828	19/64	0.316	0.278	0.031
1/2	0.5000	0.515	3/4	0.7500	0.725	1.061	0.995	21/64	0.348	0.308	0.031
5/8	0.6250	0.642	15/16	0.9375	0.906	1.326	1.244	27/64	0.444	0.400	0.062
3/4	0.7500	0.768	1 1/8	1.1250	1.088	1.591	1.494	1/2	0.524	0.476	0.062
7/8	0.8750	0.895	1 5/16	1.3125	1.269	1.856	1.742	19/32	0.620	0.568	0.062
1	1.0000	1.022	1 1/2	1.5000	1.450	2.121	1.991	21/32	0.684	0.628	0.093
1 1/8	1.1250	1.149	1 11/16	1.6875	1.631	2.386	2.239	3/4	0.780	0.720	0.093
1 1/4	1.2500	1.277	1 7/8	1.8750	1.812	2.652	2.489	27/32	0.876	0.812	0.093
1 3/8	1.3750	1.404	2 1/16	2.0625	1.994	2.917	2.738	29/32	0.940	0.872	0.093
1 1/2	1.5000	1.531	2 1/4	2.2500	2.175	3.182	2.986	1	1.036	0.964	0.093

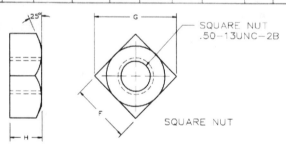

SQUARE NUT
.50−13UNC−2B

SQUARE NUT

Dimensions of Square Nuts

Nominal Size or Basic Major Dia of Thread		Width Across Flats F			Width Across Corners G		Thickness H		
		Basic	Max	Min	Max	Min	Basic	Max	Min
1/4	0.2500	7/16	0.4375	0.425	0.619	0.584	7/32	0.235	0.203
5/16	0.3125	9/16	0.5625	0.547	0.795	0.751	17/64	0.283	0.249
3/8	0.3750	5/8	0.6250	0.606	0.884	0.832	21/64	0.346	0.310
7/16	0.4375	3/4	0.7500	0.728	1.061	1.000	3/8	0.394	0.356
1/2	0.5000	13/16	0.8125	0.788	1.149	1.082	7/16	0.458	0.418
5/8	0.6250	1	1.0000	0.969	1.414	1.330	35/64	0.569	0.525
3/4	0.7500	1 1/8	1.1250	1.088	1.591	1.494	21/32	0.680	0.632
7/8	0.8750	1 5/16	1.3125	1.269	1.856	1.742	49/64	0.792	0.740
1	1.0000	1 1/2	1.5000	1.450	2.121	1.991	7/8	0.903	0.847
1 1/8	1.1250	1 11/16	1.6875	1.631	2.386	2.239	1	1.030	0.970
1 1/4	1.2500	1 7/8	1.8750	1.812	2.652	2.489	1 3/32	1.126	1.062
1 3/8	1.3750	2 1/16	2.0625	1.994	2.917	2.738	1 13/64	1.237	1.169
1 1/2	1.5000	2 1/4	2.2500	2.175	3.182	2.986	1 5/16	1.348	1.276

ASME   B 18.2.1−1996
ASME   B 18.2.2−1987 (R1993)

American Standard Hexagon Head Bolts and Nuts

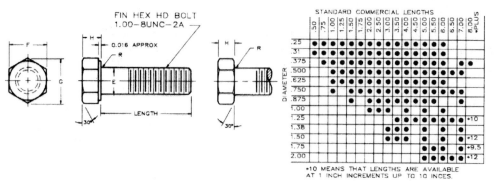

Dimensions of Hex Cap Screws (Finished Hex Bolts)

Nominal Size or Basic Product Dia		Body Dia E		Width Across Flats F			Width Across Corners G		Height H			Radius of Fillet R	
		Max	Min	Basic	Max	Min	Max	Min	Basic	Max	Min	Max	Min
1/4	0.2500	0.2500	0.2450	7/16	0.4375	0.428	0.505	0.488	5/32	0.163	0.150	0.025	0.015
5/16	0.3125	0.3125	0.3065	1/2	0.5000	0.489	0.577	0.557	13/64	0.211	0.195	0.025	0.015
3/8	0.3750	0.3750	0.3690	9/16	0.5625	0.551	0.650	0.628	15/64	0.243	0.226	0.025	0.015
7/16	0.4375	0.4375	0.4305	5/8	0.6250	0.612	0.722	0.698	9/32	0.291	0.272	0.025	0.015
1/2	0.5000	0.5000	0.4930	3/4	0.7500	0.736	0.866	0.840	5/16	0.323	0.302	0.025	0.015
9/16	0.5625	0.5625	0.5545	13/16	0.8125	0.798	0.938	0.910	23/64	0.371	0.348	0.045	0.020
5/8	0.6250	0.6250	0.6170	15/16	0.9375	0.922	1.083	1.051	25/64	0.403	0.378	0.045	0.020
3/4	0.7500	0.7500	0.7410	1 1/8	1.1250	1.100	1.299	1.254	15/32	0.483	0.455	0.045	0.020
7/8	0.8750	0.8750	0.8660	1 5/16	1.3125	1.285	1.516	1.465	35/64	0.563	0.531	0.065	0.040
1	1.0000	1.0000	0.9900	1 1/2	1.5000	1.469	1.732	1.675	39/64	0.627	0.591	0.095	0.060
1 1/8	1.1250	1.1250	1.1140	1 11/16	1.6875	1.631	1.949	1.859	11/16	0.718	0.658	0.095	0.060
1 1/4	1.2500	1.2500	1.2390	1 7/8	1.8750	1.812	2.165	2.066	25/32	0.813	0.749	0.095	0.060
1 3/8	1.3750	1.3750	1.3630	2 1/16	2.0625	1.994	2.382	2.273	27/32	0.878	0.810	0.095	0.060
1 1/2	1.5000	1.5000	1.4880	2 1/4	2.2500	2.175	2.598	2.480	15/16	0.974	0.902	0.095	0.060
1 3/4	1.7500	1.7500	1.7380	2 5/8	2.6250	2.538	3.031	2.893	1 3/32	1.134	1.054	0.095	0.060
2	2.0000	2.0000	1.9880	3	3.0000	2.900	3.464	3.306	1 7/32	1.263	1.175	0.095	0.060
2 1/4	2.2500	2.2500	2.2380	3 3/8	3.3750	3.262	3.897	3.719	1 3/8	1.423	1.327	0.095	0.060
2 1/2	2.5000	2.5000	2.4880	3 3/4	3.7500	3.625	4.330	4.133	1 17/32	1.583	1.479	0.095	0.060
2 3/4	2.7500	2.7500	2.7380	4 1/8	4.1250	3.988	4.763	4.546	1 11/16	1.744	1.632	0.095	0.060
3	3.0000	3.0000	2.9880	4 1/2	4.5000	4.350	5.196	4.959	1 7/8	1.935	1.815	0.095	0.060

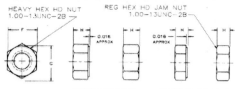

Dimensions of Hex Nuts and Hex Jam Nuts

Nominal Size or Basic Major Dia of Thread		Width Across Flats F			Width Across Corners G		Thickness Hex Nuts H			Thickness Hex Jam Nuts H		
		Basic	Max	Min	Max	Min	Basic	Max	Min	Basic	Max	Min
1/4	0.2500	7/16	0.4375	0.428	0.505	0.488	7/32	0.226	0.212	5/32	0.163	0.150
5/16	0.3125	1/2	0.5000	0.489	0.577	0.557	17/64	0.273	0.258	3/16	0.195	0.180
3/8	0.3750	9/16	0.5625	0.551	0.650	0.628	21/64	0.337	0.320	7/32	0.227	0.210
7/16	0.4375	11/16	0.6875	0.675	0.794	0.768	3/8	0.385	0.365	1/4	0.260	0.240
1/2	0.5000	3/4	0.7500	0.736	0.866	0.840	7/16	0.448	0.427	5/16	0.323	0.302
9/16	0.5625	7/8	0.8750	0.861	1.010	0.982	31/64	0.496	0.473	5/16	0.324	0.301
5/8	0.6250	15/16	0.9375	0.922	1.083	1.051	35/64	0.559	0.535	3/8	0.387	0.363
3/4	0.7500	1 1/8	1.1250	1.088	1.299	1.240	41/64	0.665	0.617	27/64	0.446	0.398
7/8	0.8750	1 5/16	1.3125	1.269	1.516	1.447	3/4	0.776	0.724	31/64	0.510	0.458
1	1.0000	1 1/2	1.5000	1.450	1.732	1.653	55/64	0.887	0.831	35/64	0.575	0.519
1 1/8	1.1250	1 11/16	1.6875	1.631	1.949	1.859	31/32	0.999	0.939	39/64	0.639	0.579
1 1/4	1.2500	1 7/8	1.8750	1.812	2.165	2.066	1 1/16	1.094	1.030	23/32	0.751	0.687
1 3/8	1.3750	2 1/16	2.0625	1.994	2.382	2.273	1 11/64	1.206	1.138	25/32	0.815	0.747
1 1/2	1.5000	2 1/4	2.2500	2.175	2.598	2.480	1 9/32	1.317	1.245	27/32	0.880	0.808

ASME B 18.2.1–1996
ASME B 18.2.2–1987 (R1993)

Fillister Head and Round Head Cap Screws

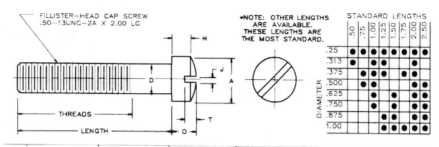

Nom-inal Size	D Body Diameter		A Head Diameter		H Height of Head		O Total Height of Head		J Width of Slot		T Depth of Slot	
	Max	Min	Max	Min	Max	Min	Max	Min	Max	Min	Max	Min
1/4	0.250	0.245	0.375	0.363	0.172	0.157	0.216	0.194	0.075	0.064	0.097	0.077
5/16	0.3125	0.307	0.437	0.424	0.203	0.186	0.253	0.230	0.084	0.072	0.115	0.090
3/8	0.375	0.369	0.562	0.547	0.250	0.229	0.314	0.284	0.094	0.081	0.142	0.112
7/16	0.4375	0.431	0.625	0.608	0.297	0.274	0.368	0.336	0.094	0.081	0.168	0.133
1/2	0.500	0.493	0.750	0.731	0.328	0.301	0.413	0.376	0.106	0.091	0.193	0.153
9/16	0.5625	0.555	0.812	0.792	0.375	0.346	0.467	0.427	0.118	0.102	0.213	0.168
5/8	0.625	0.617	0.875	0.853	0.422	0.391	0.521	0.478	0.133	0.116	0.239	0.189
3/4	0.750	0.742	1.000	0.976	0.500	0.466	0.612	0.566	0.149	0.131	0.283	0.223
7/8	0.875	0.866	1.125	1.098	0.594	0.556	0.720	0.668	0.167	0.147	0.334	0.264
1	1.000	0.990	1.312	1.282	0.656	0.612	0.803	0.743	0.188	0.166	0.371	0.291

All dimensions are given in inches.

The radius of the fillet at the base of the head:
 For sizes 1/4 to 3/8 in. incl. is 0.016 min and 0.031 max,
  7/16 to 9/16 in. incl. is 0.016 min and 0.047 max,
  5/8 to 1 in. incl. is 0.031 min and 0.062 max.

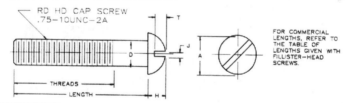

Nom-inal Size	D Body Diameter		A Head Diameter		H Height of Head		J Width of Slot		T Depth of Slot	
	Max	Min	Max	Min	Max	Min	Max	Min	Max	Min
1/4	0.250	0.245	0.437	0.418	0.191	0.175	0.075	0.064	0.117	0.097
5/16	0.3125	0.307	0.562	0.540	0.245	0.226	0.084	0.072	0.151	0.126
3/8	0.375	0.369	0.625	0.603	0.273	0.252	0.094	0.081	0.168	0.138
7/16	0.4375	0.431	0.750	0.725	0.328	0.302	0.094	0.081	0.202	0.167
1/2	0.500	0.493	0.812	0.786	0.354	0.327	0.106	0.091	0.218	0.178
9/16	0.5625	0.555	0.937	0.909	0.409	0.378	0.118	0.102	0.252	0.207
5/8	0.625	0.617	1.000	0.970	0.437	0.405	0.133	0.116	0.270	0.220
3/4	0.750	0.742	1.250	1.215	0.546	0.507	0.149	0.131	0.338	0.278

All dimensions are given in inches.

Radius of the fillet at the base of the head:
 For sizes 1/4 to 3/8 in. incl. is 0.016 min and 0.031 max,
  7/16 to 9/16 in. incl. is 0.016 min and 0.047 max,
  5/8 to 1 in. incl. is 0.031 min and 0.062 max.

ASME B 18.6.3–1972 (R1997)

Flat Head Cap Screws

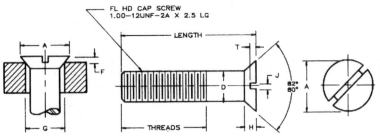

STANDARD COMMERCIAL LENGTHS

OTHER LENGTHS AND DIAMETERS ARE AVAILABLE, BUT THESE ARE THE MORE STANDARD ONES.

Nominal Size	D Body Diameter		A Head Diameter			G Gaging Diameter	H Height of Head	J Width of Slot		T Depth of Slot		F Protrusion Above Gaging Diameter	
	Max	Min	Max	Min	Absolute Min with Flat		Average	Max	Min	Max	Min	Max	Min
1/4	0.250	0.245	0.500	0.477	0.452	0.4245	0.140	0.075	0.064	0.068	0.045	0.0452	0.0307
5/16	0.3125	0.307	0.625	0.598	0.567	0.5376	0.177	0.084	0.072	0.086	0.057	0.0523	0.0354
3/8	0.375	0.369	0.750	0.720	0.682	0.6507	0.210	0.094	0.081	0.103	0.068	0.0594	0.0401
7/16	0.4375	0.431	0.8125	0.780	0.736	0.7229	0.210	0.094	0.081	0.103	0.068	0.0649	0.0448
1/2	0.500	0.493	0.875	0.841	0.791	0.7560	0.210	0.106	0.091	0.103	0.068	0.0705	0.0495
9/16	0.5625	0.555	1.000	0.962	0.906	0.8691	0.244	0.118	0.102	0.120	0.080	0.0775	0.0542
5/8	0.625	0.617	1.125	1.083	1.020	0.9822	0.281	0.133	0.116	0.137	0.091	0.0846	0.0588
3/4	0.750	0.742	1.375	1.326	1.251	1.2085	0.352	0.149	0.131	0.171	0.115	0.0987	0.0682
7/8	0.875	0.866	1.625	1.568	1.480	1.4347	0.423	0.167	0.147	0.206	0.138	0.1128	0.0776
1	1.000	0.990	1.875	1.811	1.711	1.6610	0.494	0.188	0.166	0.240	0.162	0.1270	0.0870
1 1/8	1.125	1.114	2.062	1.992	1.880	1.8262	0.529	0.196	0.178	0.257	0.173	0.1401	0.0964
1 1/4	1.250	1.239	2.312	2.235	2.110	2.0525	0.600	0.211	0.193	0.291	0.197	0.1542	0.1056
1 3/8	1.375	1.363	2.562	2.477	2.340	2.2787	0.665	0.226	0.208	0.326	0.220	0.1684	0.1151
1 1/2	1.500	1.488	2.812	2.720	2.570	2.5050	0.742	0.258	0.240	0.360	0.244	0.1825	0.1245

All dimensions are given in inches.

The maximum and minimum head diameters, A, are extended to the theoretical sharp corners.

The radius of the fillet at the base of the head shall not exceed 0.4 Max. D.

*Edge of head may be flat as shown or slightly rounded.

ASME B 18.6.3–1972 (R1997)

Machine Screws

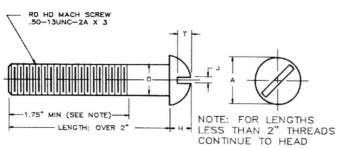

RD HD MACH SCREW
.50–13UNC–2A X 3

1.75" MIN (SEE NOTE)

LENGTH: OVER 2"

NOTE: FOR LENGTHS
LESS THAN 2" THREADS
CONTINUE TO HEAD

STANDARD LENGTHS

OTHER LENGTHS AND DIAMETERS
ARE AVAILABLE; THESE ARE THE
MORE STANDARD ONES.

Dimensions of Slotted Round Head Machine Screws

Nominal Size	D Diameter of Screw	A Head Diameter		H Head Height		J Width of Slot		T Depth of Slot	
	Basic	Max	Min	Max	Min	Max	Min	Max	Min
0	0.0600	0.113	0.099	0.053	0.043	0.023	0.016	0.039	0.029
1	0.0730	0.138	0.122	0.061	0.051	0.026	0.019	0.044	0.033
2	0.0860	0.162	0.146	0.069	0.059	0.031	0.023	0.048	0.037
3	0.0990	0.187	0.169	0.078	0.067	0.035	0.027	0.053	0.040
4	0.1120	0.211	0.193	0.086	0.075	0.039	0.031	0.058	0.044
5	0.1250	0.236	0.217	0.095	0.083	0.043	0.035	0.063	0.047
6	0.1380	0.260	0.240	0.103	0.091	0.048	0.039	0.068	0.051
8	0.1640	0.309	0.287	0.120	0.107	0.054	0.045	0.077	0.058
10	0.1900	0.359	0.334	0.137	0.123	0.060	0.050	0.087	0.065
12	0.2160	0.408	0.382	0.153	0.139	0.067	0.056	0.096	0.073
1/4	0.2500	0.472	0.443	0.175	0.160	0.075	0.064	0.109	0.082
5/16	0.3125	0.590	0.557	0.216	0.198	0.084	0.072	0.132	0.099
3/8	0.3750	0.708	0.670	0.256	0.237	0.094	0.081	0.155	0.117
7/16	0.4375	0.750	0.707	0.328	0.307	0.094	0.081	0.196	0.148
1/2	0.5000	0.813	0.766	0.355	0.332	0.106	0.091	0.211	0.159
9/16	0.5625	0.938	0.887	0.410	0.385	0.118	0.102	0.242	0.183
5/8	0.6250	1.000	0.944	0.438	0.411	0.133	0.116	0.258	0.195
3/4	0.7500	1.250	1.185	0.547	0.516	0.149	0.131	0.320	0.242

All dimensions are given in inches.

ASME B18.6.3–1972 (R1997)

American Standard Machine Screws

(The proportions of the screws can be found by multiplying the major diameter, D, by the factors given below.)

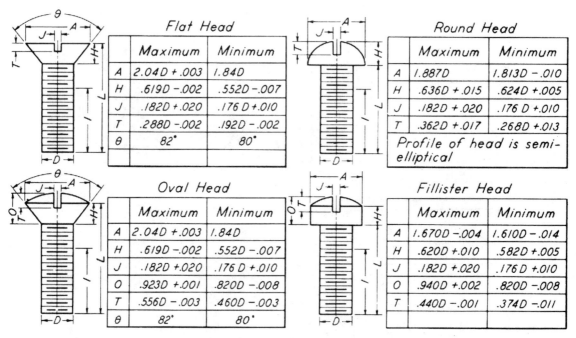

Flat Head

	Maximum	Minimum
A	2.04D + .003	1.84D
H	.619D −.002	.552D −.007
J	.182D +.020	.176 D +.010
T	.288D −.002	.192D − .002
θ	82°	80°

Round Head

	Maximum	Minimum
A	1.887D	1.813D − .010
H	.636D + .015	.624D +.005
J	.182D +.020	.176 D +.010
T	.362D +.017	.268D +.013

Profile of head is semi-elliptical

Oval Head

	Maximum	Minimum
A	2.04D +.003	1.84D
H	.619D −.002	.552D −.007
J	.182D +.020	.176 D +.010
O	.923D +.001	.820D −.008
T	.556D − .003	.460D −.003
θ	82°	80°

Fillister Head

	Maximum	Minimum
A	1.670D −.004	1.610D − .014
H	.620D +.010	.582D +.005
J	.182D +.020	.176 D +.010
O	.940D +.002	.820D −.008
T	.440D −.001	.374D −.011

ASME B 18.6.3–1972 (R1997)

Plain Washers

ASME B18.22.1–1965(R1998)

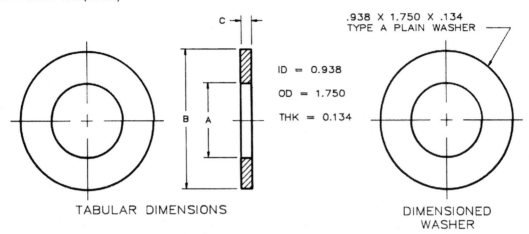

C

B   A

.938 X 1.750 X .134
TYPE A PLAIN WASHER

ID = 0.938

OD = 1.750

THK = 0.134

TABULAR DIMENSIONS

DIMENSIONED
WASHER

Dimensions of Preferred Sizes of Type A Plain Washers[a]

When specifying washers on drawings or in notes, give the inside diameter, outside diameter, and the thickness.
Example: 0.938 × 1.750 × 0.134 TYPE A PLAIN WASHER.

Nominal Washer Size[b]			Inside Diameter A			Outside Diameter B			Thickness C		
				Tolerance			Tolerance				
			Basic	Plus	Minus	Basic	Plus	Minus	Basic	Max	Min
—	—		0.078	0.000	0.005	0.188	0.000	0.005	0.020	0.025	0.016
—	—		0.094	0.000	0.005	0.250	0.000	0.005	0.020	0.025	0.016
—	—		0.125	0.008	0.005	0.312	0.008	0.005	0.032	0.040	0.025
No. 6	0.138		0.156	0.008	0.005	0.375	0.015	0.005	0.049	0.065	0.036
No. 8	0.164		0.188	0.008	0.005	0.438	0.015	0.005	0.049	0.065	0.036
No. 10	0.190		0.219	0.008	0.005	0.500	0.015	0.005	0.049	0.065	0.036
$\frac{3}{16}$	0.188		0.250	0.015	0.005	0.562	0.015	0.005	0.049	0.065	0.036
No. 12	0.216		0.250	0.015	0.005	0.562	0.015	0.005	0.065	0.080	0.051
$\frac{1}{4}$	0.250	N	0.281	0.015	0.005	0.625	0.015	0.005	0.065	0.080	0.051
$\frac{1}{4}$	0.250	W	0.312	0.015	0.005	0.734[c]	0.015	0.007	0.065	0.080	0.051
$\frac{5}{16}$	0.312	N	0.344	0.015	0.005	0.688	0.015	0.007	0.065	0.080	0.051
$\frac{5}{16}$	0.312	W	0.375	0.015	0.005	0.875	0.030	0.007	0.083	0.104	0.064
$\frac{3}{8}$	0.375	N	0.406	0.015	0.005	0.812	0.015	0.007	0.065	0.080	0.051
$\frac{3}{8}$	0.375	W	0.438	0.015	0.005	1.000	0.030	0.007	0.083	0.104	0.064
$\frac{7}{16}$	0.438	N	0.469	0.015	0.005	0.922	0.015	0.007	0.065	0.080	0.051
$\frac{7}{16}$	0.438	W	0.500	0.015	0.005	1.250	0.030	0.007	0.083	0.104	0.064
$\frac{1}{2}$	0.500	N	0.531	0.015	0.005	1.062	0.030	0.007	0.095	0.121	0.074
$\frac{1}{2}$	0.500	W	0.562	0.015	0.005	1.375	0.030	0.007	0.109	0.132	0.086

 Preferred sizes are for the most part from series previously designated "Standard Plate" and "SAE." Where common sizes existed in the two series, the SAE size is designated "N" (narrow) and the Standard Plate "W" (wide). These sizes as well as all other sizes of Type A Plain Washers are to be ordered by ID, OD, and thickness dimensions.
   [b] Nominal washer sizes are intended for use with comparable nominal screw or bolt sizes.
   [c] The 0.734 in., 1.156 in., and 1.469 in. outside diameters avoid washers which could be used in coin-operated devices.

Standard reamers are available for pins given above the line.
Pins Nos. 11 (size 0.8600), 12 (size 1.032), 13 (size 1.241), and 14 (1.523) are special sizes—hence their lengths are special. To find small diameter of pin, multiply the length by 0.02083 and substract the result from the large diameter.

ANSI B5.20–1958.

Cont.

Plain Washers (cont.)

	Nominal Washer Size[b]		Inside Diameter A			Outside Diameter B			Thickness C		
			Basic	Tolerance Plus	Tolerance Minus	Basic	Tolerance Plus	Tolerance Minus	Basic	Max	Min
$\frac{9}{16}$	0.562	N	0.594	0.015	0.005	1.156[c]	0.030	0.007	0.095	0.121	0.074
$\frac{9}{16}$	0.562	W	0.625	0.015	0.005	1.469[c]	0.030	0.007	0.109	0.132	0.086
$\frac{5}{8}$	0.625	N	0.656	0.030	0.007	1.312	0.030	0.007	0.095	0.121	0.074
$\frac{5}{8}$	0.625	W	0.688	0.030	0.007	1.750	0.030	0.007	0.134	0.160	0.108
$\frac{3}{4}$	0.750	N	0.812	0.030	0.007	1.469	0.030	0.007	0.134	0.160	0.108
$\frac{3}{4}$	0.750	W	0.812	0.030	0.007	2.000	0.030	0.007	0.148	0.177	0.122
$\frac{7}{8}$	0.875	N	0.938	0.030	0.007	1.750	0.030	0.007	0.134	0.160	0.108
$\frac{7}{8}$	0.875	W	0.938	0.030	0.007	2.250	0.030	0.007	0.165	0.192	0.136
1	1.000	N	1.062	0.030	0.007	2.000	0.030	0.007	0.134	0.160	0.108
1	1.000	W	1.062	0.030	0.007	2.500	0.030	0.007	0.165	0.192	0.136
$1\frac{1}{8}$	1.125	N	1.250	0.030	0.007	2.250	0.030	0.007	0.134	0.160	0.108
$1\frac{1}{8}$	1.125	W	1.250	0.030	0.007	2.750	0.030	0.007	0.165	0.192	0.136
$1\frac{1}{4}$	1.250	N	1.375	0.030	0.007	2.500	0.030	0.007	0.165	0.192	0.136
$1\frac{1}{4}$	1.250	W	1.375	0.030	0.007	3.000	0.030	0.007	0.165	0.192	0.136
$1\frac{3}{8}$	1.375	N	1.500	0.030	0.007	2.750	0.030	0.007	0.165	0.192	0.136
$1\frac{3}{8}$	1.375	W	1.500	0.045	0.010	3.250	0.045	0.010	0.180	0.213	0.153
$1\frac{1}{2}$	1.500	N	1.625	0.030	0.007	3.000	0.030	0.007	0.165	0.192	0.136
$1\frac{1}{2}$	1.500	W	1.625	0.045	0.010	3.500	0.045	0.010	0.180	0.213	0.153
$1\frac{5}{8}$	1.625		1.750	0.045	0.010	3.750	0.045	0.010	0.180	0.213	0.153
$1\frac{3}{4}$	1.750		1.875	0.045	0.010	4.000	0.045	0.010	0.180	0.213	0.153
$1\frac{7}{8}$	1.875		2.000	0.045	0.010	4.250	0.045	0.010	0.180	0.213	0.153
2	2.000		2.125	0.045	0.010	4.500	0.045	0.010	0.180	0.213	0.153
$2\frac{1}{4}$	2.250		2.375	0.045	0.010	4.750	0.045	0.010	0.220	0.248	0.193
$2\frac{1}{2}$	2.500		2.625	0.045	0.010	5.000	0.045	0.010	0.238	0.280	0.210
$2\frac{3}{4}$	2.750		2.875	0.065	0.010	5.250	0.065	0.010	0.259	0.310	0.228
3	3.000		3.125	0.065	0.010	5.500	0.065	0.010	0.284	0.327	0.249

ASME B18.22.1–1965 (R1998)

Scales

Scales

When you create a drawing with a computer, it is drawn full size, but it must be plotted on a piece of paper that may not be large enough to draw full-size views of small objects. (Engineering drawing paper comes in standard sizes.) When the drawing is plotted, it must be drawn smaller than full size and it must be drawn to some scale. The scale must always be stated on the drawing.

Drawing scales used in engineering are not chosen randomly. They are standardized. The type of scale you use depends upon what you are drawing. The common types of scales are:

- an architect's scale
- an engineer's scale (sometimes called a civil engineer's scale, although its use is not confined to civil engineering)
- a mechanical engineer's scale.
  Metric scales are discussed in Chapter 4.

Architect's Scale

An architect's scale is divided into sections representing feet and inches. You are probably familiar with a full-size scale where a foot is divided into 12 inches and inches are divided into sixteenths. This full-size scale is suitable for drawing small objects, but could not be used to draw a building.

Standard drawing scales are:

2/32"	= 1' - 0" (1:128)	(one inch represents 128 inches)
3/16"	= 1' - 0" (1:48)	
1/8"	= 1' - 0" (1:96)	
3/8"	= 1' - 0" (1:32)	
1/2"	= 1' - 0" (1:24)	
3/4"	= 1' - 0" (1:16)	
1"	= 1' - 0" (1:12)	
1 1/2"	= 1' - 0" (1:8)	
3"	= 1' - 0" (1:4)	
Full size	= 1' - 0" (1:1)	divided into sixteenths

Figure E.1 illustrates a 1' = 1' - 0" scale.

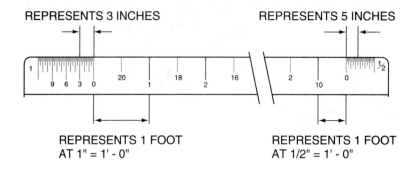

Figure E.1   Architect's scale 1" = 1' - 0"

A one-inch length on the scale represents one foot in space. The first one-inch division is divided into 12 parts, each representing one inch.

A line 1' - 7" long would be laid out starting from the right end and measuring to the left, as shown in Figure E.2. One inch on the drawing represents one foot on the real object.

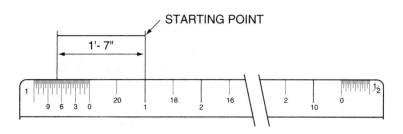

Figure E.2   Laying out a length using an architect's scale

A larger reduction would be used to lay out longer lengths. Figure E.3 shows a length of 4' - 8" laid out at a scale of 1/2" = 1' - 0".

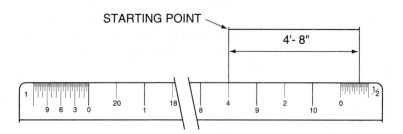

Figure E.3   Using a 1/2" = 1' - 0" scale

This length is laid out from right to left. One half-inch on the drawing represents one foot on the object.

Engineer's Scale (Civil Engineer's Scale)

The major divisions on an engineer's scale are one inch, but each scale has a different number of divisions per inch.

The scales are:

10 10 divisions per inch
20 20 divisions per inch
30 30 divisions per inch
40 40 divisions per inch
50 50 divisions per inch
60 60 divisions per inch

Figure E.4 shows a 10 scale.

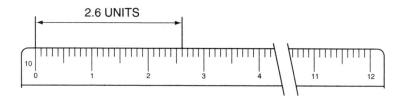

Figure E.4   Engineer's 10 scale

A length of 2.6 units is shown. The units could be feet (1" = 10 ft), miles (1" = 10 miles), or some other quantity, 1" = 100 N (using the 10 scale), or 1" = 500 pounds (using the 50 scale). A length of 19 units is laid out in Figure E.5 using a 50 scale (1" = 50 units). The units could be feet, miles, or another quantity.

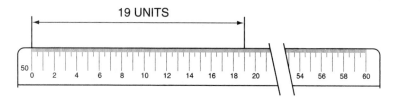

Figure E.5   Laying out a length using an engineer's scale

Mechanical Engineer's Scale

The units on a mechanical engineer's scale are inches and common fractions of an inch (1/4, 1/16). The scale is used to lay out lengths at half-size (one-half full size, 1:2), quarter-size (one-quarter full size, 1:4), or one-eighth scale (one-eight full size, 1:8). A full-size drawing would use the familiar inch scale, divided into sixteenths. A half-size mechanical engineer's scale is shown in Figure E.6. At this scale, one inch on the drawing represents two inches on the object.

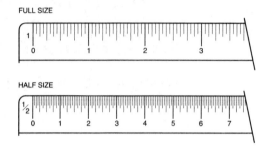

Figure E.6   Mechanical engineer's scale used to draw at half-size

Appendix F

Table of Notations for Common Structural Steel Shapes

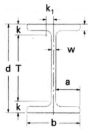

Table of Notations For Common Structural Steel Shapes

W SHAPES
W690 - W610

PROPERTIES

DIMENSIONS AND SURFACE AREAS

Designation[1]	Nominal Mass	Theo-retical Mass	Depth d	Flange Width b	Flange Thick-ness t	Web Thick-ness w	Distances					Surface Area (m²) per metre of length		Imperial Designation
							a	T	k	k₁	d-2t	Total	Minus Top of Top Flange	
	kg/m	kg/m	mm	mm	mm	mm	mm	mm	mm	mm	mm			
W690														
x802	802	801.4	826	387	89.9	50.0	169	603	111	45	646	3.10	2.71	W27x539
x548	548	547.5	772	372	63.0	35.1	168	603	85	38	646	2.96	2.59	W27x368
x500	500	499.3	762	369	57.9	32.0	169	603	79	36	646	2.94	2.57	W27x336
x457	457	457.0	752	367	53.1	29.5	169	603	75	35	646	2.91	2.55	W27x307
x419	419	418.0	744	364	49.0	26.9	169	603	71	33	646	2.89	2.53	W27x281
x384*	384	383.5	736	362	45.0	24.9	169	603	67	32	646	2.87	2.51	W27x258
x350	350	349.9	728	360	40.9	23.1	168	603	62	32	646	2.85	2.49	W27x235
x323	323	323.2	722	359	38.1	21.1	169	603	60	31	646	2.84	2.48	W27x217
x289	289	287.9	714	356	34.0	19.0	169	603	56	30	646	2.81	2.46	W27x194
x265	265	264.5	706	358	30.2	18.4	170	603	52	29	646	2.81	2.45	W27x178
x240	240	239.9	701	356	27.4	16.8	170	603	49	28	646	2.79	2.44	W27x161
x217	217	217.8	695	355	24.8	15.4	170	602	46	28	645	2.78	2.42	W27x146
W690														
x192	192	191.4	702	254	27.9	15.5	119	603	49	28	646	2.39	2.14	W27x129
x170	170	169.9	693	256	23.6	14.5	121	603	45	27	646	2.38	2.13	W27x114
x152	152	152.1	688	254	21.1	13.1	120	603	43	27	646	2.37	2.11	W27x102
x140	140	139.8	684	254	18.9	12.4	121	603	40	26	646	2.36	2.11	W27x94
x125	125	125.5	678	253	16.3	11.7	121	602	38	26	645	2.34	2.09	W27x84
W610														
x551	551	551.2	711	347	69.1	38.6	154	530	91	39	573	2.73	2.39	W24x370
x498	498	498.3	699	343	63.0	35.1	154	530	85	38	573	2.70	2.36	W24x335
x455	455	454.2	689	340	57.9	32.0	154	530	79	36	573	2.67	2.33	W24x306
x415	415	415.6	679	338	53.1	29.5	154	530	75	35	573	2.65	2.31	W24x279
x372	372	372.3	669	335	48.0	26.4	154	530	70	33	573	2.63	2.29	W24x250
x341	341	340.4	661	333	43.9	24.4	154	530	65	32	573	2.61	2.27	W24x229
x307	307	307.3	653	330	39.9	22.1	154	530	61	31	573	2.58	2.25	W24x207
x285	285	285.4	647	329	37.1	20.6	154	530	59	30	573	2.57	2.24	W24x192
x262	262	261.2	641	327	34.0	19.0	154	530	56	30	573	2.55	2.23	W24x176
x241	241	241.7	635	329	31.0	17.9	156	530	53	29	573	2.55	2.22	W24x162
x217	217	217.9	628	328	27.7	16.5	156	530	49	28	573	2.54	2.21	W24x146
x195	195	195.6	622	327	24.4	15.4	156	530	46	28	573	2.52	2.19	W24x131
x174	174	174.3	616	325	21.6	14.0	156	530	43	27	573	2.50	2.18	W24x117
x155	155	154.9	611	324	19.0	12.7	156	530	41	26	573	2.49	2.17	W24x104
W610														
x153	153	153.6	623	229	24.9	14.0	108	530	46	27	573	2.13	1.91	W24x103
x140	140	140.1	617	230	22.2	13.1	108	530	44	27	573	2.13	1.90	W24x94
x125	125	125.1	612	229	19.6	11.9	109	530	41	26	573	2.12	1.89	W24x84
x113	113	113.4	608	228	17.3	11.2	108	530	39	26	573	2.11	1.88	W24x76
x101	101	101.7	603	228	14.9	10.5	109	530	36	25	573	2.10	1.87	W24x68
x91	91	90.9	598	227	12.7	9.7	109	530	34	25	573	2.08	1.86	W24x61
x84	84	84.0	596	226	11.7	9.0	109	530	33	25	573	2.08	1.85	W24x56

[Adapted from *Handbook of Steel Construction*, 7th Edition. Canadian Institute of Steel Construction, 1997, p.6-45.]

Common Pipe Sizes and Symbols

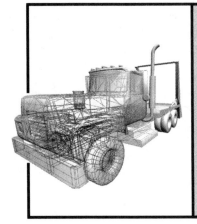

Standard Pipe[a, b]

Welded and Seamless Wrought Steel Pipe

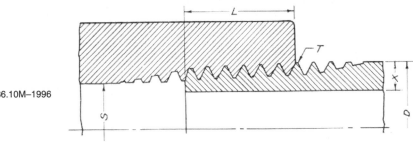

ASME B36.10M–1996

[a] Stainless Steel Pipe, ASME B36.19–M–1985 (R1994)
[b] A pipe size may be designated by giving the nominal pipe size and wall thickness or by giving the nominal pipe size and weight per linear foot.
[c] Refers to ANSI Standard schedule number, approximate values for the expression 1,000 × P/S. Schedule 40–standard weight.
[d] Schedule 80–extra strong.
[e] Not ANSI Standard, but commercially available in both wrought iron and steel.
[f] Plain ends.

[Source: French, T.E., and C.J. Vierck. *Engineering Drawing and Graphic Technology*, 12th Edition. New York: McGraw Hill, 1978, p. A69.]

Nominal pipe size	Actual outside diam. D	Tap-drill size S	Thd. /in. T	Distance pipe enters fittings L	Wall thickness X			Weight, lb/ft[f]		
					Standard 40[c]	Extra strong 80[d]	Double extra strong[e]	Standard 40[c]	Extra strong 80[d]	Double extra strong[e]
1/8	0.405	1 1/32	27	3/16	0.070	0.098	. . . .	0.25	0.32	
1/4	0.540	7/16	18	9/32	0.090	0.122	. . . .	0.43	0.54	
3/8	0.675	37/64	18	19/64	0.093	0.129	. . . .	0.57	0.74	
1/2	0.840	23/32	14	3/8	0.111	0.151	0.307	0.86	1.09	1.714
3/4	1.050	59/64	14	13/32	0.115	0.157	0.318	1.14	1.48	2.440
1	1.315	1 5/32	11½	1/2	0.136	0.183	0.369	1.68	2.18	3.659
1¼	1.660	1 1/2	11½	35/64	0.143	0.195	0.393	2.28	3.00	5.214
1½	1.900	1 47/64	11½	9/16	0.148	0.204	0.411	2.72	3.64	6.408
2	2.375	2 7/32	11½	37/64	0.158	0.223	0.447	3.66	5.03	9.029
2½	2.875	2 5/8	8	7/8	0.208	0.282	0.565	5.80	7.67	13.695
3	3.5	3 1/4	8	15/16	0.221	0.306	0.615	7.58	10.3	18.583
3½	4.0	3 3/4	8	1	0.231	0.325	. . . .	9.11	12.5	
4	4.5	4 1/4	8	1 1/16	0.242	0.344	0.690	10.8	15.0	27.451
5	5.563	5 5/16	8	1 5/32	0.263	0.383	0.768	14.7	20.8	38.552
6	6.625	6 5/16	8	1 1/4	0.286	0.441	0.884	19.0	28.6	53.160
8	8.625	. . . . .	8	1 15/32	0.329	0.510	0.895	28.6	43.4	72.424
10	10.75	. . . . .	8	1 43/64	0.372	0.606	. . . .	40.5	64.4	
12	12.75	. . . . .	8	1 7/8	0.414	0.702	. . . .	53.6	88.6	
14 OD	14.0	. . . . .	8	2	0.437	0.750	. . . .	62.2	104.	
16 OD	16.0	. . . . .	8	2 13/64	0.500	. . . .	. . . .	82.2		
18 OD	18.0	. . . . .	8	2 13/32	0.562	. . . .	. . . .	103.		
20 OD	20.0	. . . . .	8	2 19/32	0.562	. . . .	. . . .	115.		
24 OD	24.0	. . . . .	8	3						

Graphic Symbols for Piping and Heating

PIPING

Piping, General	(Lettered with name of material conveyed)
Non-intersecting Pipes	

(To differentiate lines of piping on a drawing the following symbols may be used.)

Air	Cold Water	Steam
Gas	Hot Water	Condensate
Oil	Vacuum	Refrigerant

PIPE FITTINGS AND VALVES

	Flanged	Screwed	Bell and Spigot	Welded	Soldered
Joint					
Elbow—90 deg					
Elbow—45 deg					
Elbow—Turned Up					
Elbow—Turned Down					
Elbow—Long Radius					
Side Outlet Elbow Outlet Down					
Side Outlet Elbow Outlet Up					
Base Elbow					
Double Branch Elbow					
Reducing Elbow					
Reducer					
Eccentric Reducer					
Tee-Outlet Up					
Tee-Outlet Down					
Tee					
Side Outlet Tee Outlet Up					
Side Outlet Tee Outlet Down					
Single Sweep Tee					
Double Sweep Tee					
Cross					
Lateral					
Gate Valve					

ASME   Y32.2.3–1949 (R1994)
Graphic symbols for pipe fittings, valves, and pipes

Cont.

Graphic Symbols for Piping and Heating (cont.)

PIPING

	Flanged	Screwed	Bell and Spigot	Welded	Soldered
Globe Valve					
Angle Globe Valve					
Angle Gate Valve					
Check Valve					
Angle Check Valve					
Stop Cock					
Safety Valve					
Quick Opening Valve					
Float Operating Valve					
Motor Operated Gate Valve					
Motor Operated Globe Valve					
Expansion Joint Flanged					
Reducing Flange					
Union	(See Joint)				
Sleeve					
Bushing					

HEATING AND VENTILATING

Lock and Shield Valve		Tube Radiator	(Plan) (Elev.)	Exhaust Duct, Section	
Reducing Valve		Wall Radiator	(Plan) (Elev.)	Butterfly Damper	(Plan or Elev.) (Elev. or Plan)
Diaphragm Valve		Pipe Coil	(Plan) (Elev.)	Deflecting Damper Rectangular Pipe	
Thermostat	(T)	Indirect Radiator	(Plan) (Elev.)	Vanes	
Radiator Trap	(Plan) (Elev.)	Supply Duct, Section		Air Supply Outlet	
				Exhaust Inlet	

HEAT-POWER APPARATUS

Flue Gas Reheater (Intermediate Superheater)		Steam Turbine		Automatic By pass Valve	
Steam Generator (Boiler)		Condensing Turbine		Automatic Valve Operated by Governor	
Live Steam Superheater		Open Tank		Pumps Air Service Boiler Feed Condensate Circulating Water Reciprocating	
Feed Heater With Air Outlet		Closed Tank			
Surface Condenser		Automatic Reducing Valve		Dynamic Pump (Air Ejector)	

ASME Y32.2.4–1949 (R1998)
Graphic symbols for heating, ventilating, and air conditioning

ASME Y32.2.6–1950 (R1998)
Graphic symbols for heat-power apparatus

[Source: French, T.E., and C.J. Vierck. *Engineering Drawing and Graphic Technology*, 12th Edition. New York: McGraw-Hill, 1978, p.A83–A84.]

Appendix H

Graphic Symbols for Electrical and Electronic Diagrams

ANSI Standard Graphic Symbols for Electrical Diagrams

Single-line symbols are shown at the left, complete symbols at the right, and symbols for both purposes are centred in each column.
Listing is alphabetical

ADJUSTABLE
CONTINUOUSLY ADJUSTABLE (Variable)
The shaft of the arrow is drawn at about 45 degrees across the body of the symbol.

AMPLIFIER
See also MACHINE, ROTATING

General
The triangle is pointed in the direction of transmission.

Amplifier type may be indicated in the triangle by words, standard abbreviations, or a letter combination from the following list.

BDG	Bridging	MON	Monitoring
BST	Booster	PGM	Program
CMP	Compression	PRE	Preliminary
DC	Direct Current	PWR	Power
EXP	Expansion	TRQ	Torque
LIM	Limiting		

Applications

Booster amplifier with two inputs

Monitoring amplifier with two outputs

Amplifier with associated power supply

ANTENNA

General
Types or functions may be indicated by words or abbreviations adjacent to the symbol.

Dipole

ARRESTER (Electric Surge, Lightning, etc.)
GAP

General

Carbon block
The sides of the rectangle are to be approximately in the ratio of 1 to 2 and the space between rectangles shall be approximately equal to the width of a rectangle.

Electrolytic or aluminum cell
This symbol is not composed of arrowheads.

Protective gap
These arrowheads shall not be filled.

Sphere gap

Multigap, general

ATTENUATOR
See also PAD

General

Balanced, general

Unbalanced, general

BATTERY
The long line is always positive, but polarity may be indicated in addition.
Example:

Generalized direct-current source

BREAKER, CIRCUIT
If it is desired to show the condition causing the breaker to trip, the relay-protective-function symbols   may be used alongside the breaker symbol.

General
Note 1—Use appropriate number of single-line diagram symbols.

SEE NOTE 1

Air or, if distinction is needed, for alternating-current circuit breaker rated at 1,500 volts or less and for direct-current circuit breaker.

SEE NOTE 1

CAPACITOR
See also TERMINATION

General
If it is necessary to identify the capacitor electrodes, the curved element shall represent the outside electrode in fixed paper-dielectric and ceramic-dielectric capacitors, the negative electrode in electrolytic capacitors, the moving element in adjustable and variable capacitors, and the low-potential element in feed-through capacitors.

Application: shielded capacitor

Application: adjustable or variable capacitor

If it is necessary to identify trimmer capacitors, the letter T should appear adjacent to the symbol.

Application: adjustable or variable capacitors with mechanical linkage of units

Shunt capacitor

Feed-through capacitor (with terminals shown on feed-through element)
Commonly used for bypassing high-frequency currents to chassis.

Application: feed-through capacitor between 2 inductors with third lead connected to chassis

CELL, PHOTOSENSITIVE (Semiconductor)

λ indicates that the primary characteristic of the element within the circle is designed to vary under the influence of light.

Asymmetrical photoconductive transducer (resistive)
This arrowhead shall be solid.

Symmetrical photoconductive transducer; selenium cell

CHASSIS
FRAME
See also GROUND
The chassis or frame is not necessarily at ground potential.

COIL, BLOWOUT

COIL, OPERATING
See also INDUCTOR; WINDING
Note 3—The asterisk is not a part of the symbol. Always replace the asterisk by a device designation.

OR   OR

* SEE NOTE 3

CONNECTION, MECHANICAL –
MECHANICAL INTERLOCK
The preferred location of the mechanical connection is as shown in the various applications, but other locations may be equally acceptable.

Mechanical connection (*short dashes*)

Cont.

ANSI Standard Graphic Symbols for Electrical Diagrams (cont.)

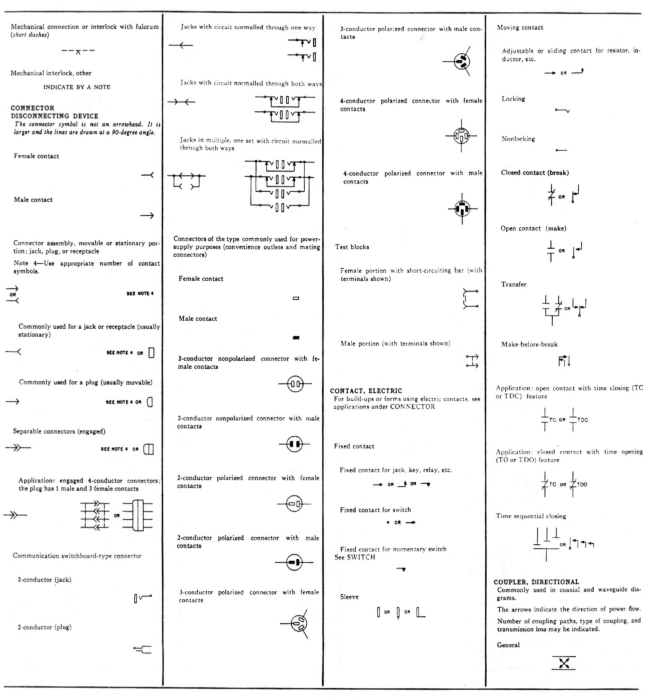

Mechanical connection or interlock with fulcrum (*short dashes*)

Mechanical interlock, other

INDICATE BY A NOTE

CONNECTOR
DISCONNECTING DEVICE
The connector symbol is not an arrowhead. It is larger and the lines are drawn at a 90-degree angle.

Female contact

Male contact

Connector assembly, movable or stationary portion; jack, plug, or receptacle

Note 4—Use appropriate number of contact symbols.

SEE NOTE 4

Commonly used for a jack or receptacle (usually stationary)

SEE NOTE 4 OR

Commonly used for a plug (usually movable)

SEE NOTE 4 OR

Separable connectors (engaged)

SEE NOTE 4 OR

Application: engaged 4-conductor connectors; the plug has 1 male and 3 female contacts

OR

Communication switchboard-type connector

2-conductor (jack)

2-conductor (plug)

Jacks with circuit normalled through one way

Jacks with circuit normalled through both ways

Jacks in multiple, one set with circuit normalled through both ways

Connectors of the type commonly used for power-supply purposes (convenience outlets and mating connectors)

Female contact

Male contact

2-conductor nonpolarized connector with female contacts

2-conductor nonpolarized connector with male contacts

2-conductor polarized connector with female contacts

2-conductor polarized connector with male contacts

3-conductor polarized connector with female contacts

3-conductor polarized connector with male contacts

4-conductor polarized connector with female contacts

4-conductor polarized connector with male contacts

Test blocks

Female portion with short-circuiting bar (with terminals shown)

Male portion (with terminals shown)

CONTACT, ELECTRIC
For build-ups or forms using electric contacts, see applications under CONNECTOR

Fixed contact

Fixed contact for jack, key, relay, etc.

OR   OR

Fixed contact for switch

OR

Fixed contact for momentary switch
See SWITCH

Sleeve

OR   OR

Moving contact

Adjustable or sliding contact for resistor, inductor, etc.

OR

Locking

Nonlocking

Closed contact (break)

OR

Open contact (make)

OR

Transfer

OR

Make-before-break

Application: open contact with time closing (TC or TDC) feature

TC OR TDC

Application: closed contact with time opening (TO or TDO) feature

TO OR TDO

Time sequential closing

OR

COUPLER, DIRECTIONAL
Commonly used in coaxial and waveguide diagrams.

The arrows indicate the direction of power flow.

Number of coupling paths, type of coupling, and transmission loss may be indicated.

General

Cont.

ANSI Standard Graphic Symbols for Electrical Diagrams (cont.)

Applications

E-plane aperture coupling, 30-decibel transmission loss

Loop coupling, 30-decibel transmission loss

Probe coupling, 30-decibel transmission loss

Resistance coupling, 30-decibel transmission loss

DIRECTION OF FLOW OF POWER, SIGNAL, OR INFORMATION

One-way
Note 10—The lower symbol is used if it is necessary to conserve space. *The arrowhead in the lower symbol shall be filled.*

SEE NOTE 10

Both ways

SEE NOTE 10

DISCONTINUITY
A component that exhibits throughout the frequency range of interest the properties of the type of circuit element indicated by the symbol within the triangle.

Commonly used for coaxial and waveguide transmission.

Equivalent series element, general

Capacitive reactance

ELEMENT, CIRCUIT (General)
Note 12—The asterisk is not a part of the symbol. Always indicate the type of apparatus by appropriate words or letters in the rectangle.

* SEE NOTE 12

Accepted abbreviations in the latest edition of American Standard Z32.13 may be used in the rectangle.

The following letter combinations may be used in the rectangle.

CB	Circuit breaker	NET	Network
DIAL	Telephone dial	PS	Power supply
EQ	Equalizer	RU	Reproducing unit
FAX	Facsimile set		
FL	Filter	RG	Recording unit
FL-BE	Filter, band elimination	TEL	Telephone station
FL-BP	Filter, band pass	TPR	Teleprinter
FL-HP	Filter, high pass	TTY	Teletypewriter
FL-LP	Filter, low pass		

ELEMENT, THERMAL
Thermomechanical transducer

Note 13—Use appropriate number of single-line diagram symbols.

SEE NOTE 13

FUSE

SEE NOTE 14

Note 14—Use appropriate number of single-line diagram symbols.

Fusible element

SEE NOTE 14

GROUND
See also CHASSIS; FRAME

INDUCTOR
WINDING
General

Either symbol may be used in the following subparagraphs.

OR

If it is desired especially to distinguish magnetic-core inductors

LAMP

Ballast lamp; ballast tube

The primary characteristic of the element within the circle is designed to vary nonlinearly with the temperature of the element.

Fluorescent lamp
2-terminal

4-terminal

Glow lamp; cold-cathode lamp; neon lamp

Alternating-current type

Direct-current type
See also TUBE, ELECTRON

Incandescent-filament illuminating lamp

MACHINE, ROTATING

Basic

Generator, general

Motor, general

Motor, multispeed

USE BASIC MOTOR SYMBOL AND NOTE SPEEDS

Rotating armature with commutator and brushes

Wound rotor

Field, generator or motor

Compensating or commutating

Series

Shunt, or separately excited

Permanent magnet

Winding symbols

Motor and generator winding symbols may be shown in the basic circle using the following representations.

1-phase

2-phase

3-phase wye (ungrounded)

3-phase wye (grounded)

3-phase delta

6-phase diametrical

6-phase double-delta

MOTION, MECHANICAL

Translation, one direction

Translation, both directions

Rotation, one direction

Rotation, both directions

Cont.

ANSI Standard Graphic Symbols for Electrical Diagrams (cont.)

NETWORK

General

NET

OSCILLATOR GENERALIZED ALTERNATING-CURRENT SOURCE

PAD

See also ATTENUATOR

General

Balanced, general

Unbalanced, general

PATH, TRANSMISSION

Air or space path

Dielectric path other than air

Commonly used for coaxial and waveguide transmission.

DIEL

Crossing of paths or conductors not connected
The crossing is not necessarily at a 90-degree angle.

Junction of paths or conductors

Junction (if desired)

Application: junction of different-size cables

Junction of connected paths, conductors, or wires

OR

OR ONLY IF REQUIRED
BY SPACE LIMITATION

RELAY

See also CONTACT

Fundamental symbols for contacts, mechanical connections, coils, etc., are the basis of relay symbols and should be used to represent relays on complete diagrams.

Basic

Relay coil

Note 21—The asterisk is not a part of the symbol. Always replace the asterisk by a device designation.

⊙ OR ⋛ OR

＊ SEE NOTE 21

Application: 2-pole double-make

RESISTOR

Note 22—The asterisk is not a part of the symbol. Always add identification within or adjacent to the rectangle.

General

—WW— OR

＊ SEE NOTE 22

Tapped resistor

OR

＊ SEE NOTE 22

Application: with adjustable contact

OR

＊ SEE NOTE 22

Application: adjustable or continuously adjustable (variable) resistor

OR

＊ SEE NOTE 22

Heating resistor

—WW— OR

＊ SEE NOTE 22

SWITCH

Fundamental symbols for contacts, mechanical connections, etc., may be used for switch symbols.

The standard method of showing switches is in a position with no operating force applied. For switches that may be in any one of two or more positions with no operating force applied and for switches actuated by some mechanical device (as in air-pressure, liquid-level, rate-of-flow, etc., switches), a clarifying note may be necessary to explain the point at which the switch functions.

Single throw, general

Double throw, general

Application: 2-pole double-throw switch with terminals shown

Knife switch, general

Switch, nonlocking; momentary or spring return

The symbols to the left are commonly used for spring buildups in key switches, relays, and jacks.
The symbols to the right are commonly used for toggle switches.

Circuit closing (make)

OR

Circuit opening (break)

OR

Switch, locking

The symbols to the left are commonly used for spring buildups in key switches, relays, and jacks.
The symbols to the right are commonly used for toggle switches.

Circuit closing (make)

OR

Circuit opening (break)

OR

Transfer, 2-position

OR

TERMINAL, CIRCUIT

Terminal board or terminal strip with 4 terminals shown; group of 4 terminals

Number and arrangement as convenient.

TRANSFORMER

General

Either winding symbol may be used.

Additional windings may be shown or indicated by a note.

For power transformers, use polarity marking H_1-X_1, etc., from American Standard C6.1.

In coaxial and waveguide circuits, this symbol will represent a taper or step transformer without mode change.

OR

If it is desired especially to distinguish a magnetic-core transformer

TUBE, ELECTRON

Tube-component symbols are shown first. These are followed by typical applications showing the use of these specific symbols in the various classes of devices such as thermionic, cold-cathode, and photoemissive tubes of varying structures and combinations of elements (triodes, pentodes, cathode-ray tubes, magnetrons, etc.).

Lines outside of the envelope are not part of the symbol but are electrical connections thereto.

Connections between the external circuit and electron tube symbols within the envelope may be located as required to simplify the diagram.

Emitting electrode

Directly heated (filamentary) cathode
Note—Leads may be connected in any convenient manner to ends of the ∧ provided the identity of the ∧ is retained.

Indirectly heated cathode
Lead may be connected to either extreme end of the ⌐ or, if required, to both ends, in any convenient manner.

Cold cathode (including ionically heated cathode)

Photocathode

Pool cathode

Ionically heated cathode with provision for supplementary heating

R

[Source: IEEE 315–1975 Graphic Symbols for Electrical and Electronic Diagrams]

Appendix I
SI Units and Conversion Factors

Complete instructions for the use of SI units and conversion factors for other quantities can be found in:

CSA-Z234.1-89	Canadian Metric Practice Guide
ASME S1-1	ASME Orientation Guide for the Use of SI (Metric) Units
ASTM/IEEE SI10-1977	Standard for the Use of the International System of Units (SI)
ISO 1000:1992	SI Units and Recommendations for the Use of Their Multiples and of Certain Other Units plus standards relating to specific industries.

Additional information can be found at:

- www.cssinfo.com
- www.asme.org
- www.ansi.org

Base Units

Unit	Name	Symbol
Length	metre (meter)	m
Mass	kilogram	kg
Time	second	s
Electric current	ampere	A
Thermodynamic temperature	kelvin (not degree kelvin)	K
Amount of substance	mole	mol
Luminous intensity	candela	cd

Derived Units With Special Names

Unit	Name	Symbol	Derived From
Frequency	hertz	Hz	s^{-1}
Force	newton	N	$m \cdot kg/s^2$
Pressure	pascal	Pa	N/m^2
Energy, work, quantity of heat	joule	J	$N \cdot m$
Quantity of electricity	watt	W	J/s
Celsius temperature	degree Celsius	°C	

SI Prefixes

Prefix	Multiplying Factor	Symbol
exa	10^{18}	E
peta	10^{15}	P
tera	10^{12}	T
giga	10^{9}	G
mega	10^{6}	M
kilo	10^{3}	k
hecto	10^{2}	h
deca	10^{1}	da
deci	10^{-1}	d
centi	10^{-2}	c
milli	10^{-3}	m
micro	10^{-6}	μ
nano	10^{-9}	n
pico	10^{-12}	p
femto	10^{-15}	f
atto	10^{-18}	a

Common Conversion Factors

The values in the following table are rounded.

Item	Conversion Factor	
Mass and Density	1 pound per cubic foot	= 16.018 kg/m³
	1 pound per gallon (US)	= 99.776 kg/m³
	1 pound per gallon (UK)	= 119.826 kg/m³
Energy and Power	1 Btu (mean)	= 1.056 kJ
	1 calorie	= 4.187 J
	1 foot pound force	= 1.356 J
	1 horsepower hour	= 2.685 MJ
	1 horsepower	= 746 W
	1 kilowatt hour	= 3.6 MJ
Force	1 pound force	= 4.448 N
	1 kip (1000 pounds force)	= 4.448 kN
	1 kilogram force	= 9.807 N
Heat		
Conductivity	1 Btu/h·ft·°F	= 1.731 W/m·K
Specific heat	1 Btu/pound mass·°F	= 4187 J/kg·K
Length	1 foot	= 0.3048 m
	1 inch	= 25.4 mm
	1 micron	= 1 μm
	1 mile	= 1.609 km
	1 yard	= 0.9144 m
	1 nautical mile (US)	= 1.852 km
Mass	1 ounce (avoirdupois)	= 28.349 g
	1 pound (avoirdupois)	= 0.454 kg
	1 slug	= 14.594 kg
	1 ton (2000 pounds)	= 0.907 Mg
Pressure (force per unit area)	1 bar	= 100 kPa
	1 inch of water (68°F, 20°C)	= 248.843 Pa
	1 inch of mercury (68°F, 20°C)	= 3.374 kPa
	1 pound force per square inch (psi)	= 6.895 kPa
	1 pound force per square foot	= 47.880 Pa
	1 kip per square inch (ksi)	= 6.895 Mpa

Item	Conversion Factor	
Temperature	Fahrenheit temperature	= 1.8 (Celsius temperature) + 32
Torque	1 foot pound force	= 1.356 N·m
	1 inch pound force	= 0.113 N·m
Velocity	1 foot per minute	= 5.08 mm/s
	1 foot per second	= 304.8 mm/s
	1 mile per hour	= 1.609 km/h
	1 mile per minute	= 26.822 m/s
	1 knot (UK)	= 1.853 km/h
Volume	1 litre (liter)	= 1 dm^3
	1 cubic inch	= 16.387 cm^3
	1 cubic foot	= 28.317 dm^3
	1 gallon (US)	= 3.785 dm^3
	1 gallon (UK)	= 4.546 dm^3
Volume Rate of Flow	1 cubic foot per minute	= 0.472 dm^3/s
	1 cubic foot per second	= 28.317 dm^3/s
	1 gallon (US) per minute	= 63.090 cm^3/s
	1 gallon (UK) per minute	= 75.768 cm^3/s

Index

The boldface page numbers indicate definitions.